Sustained Simulation Performance 2018 and 2019

Michael M. Resch • Yevgeniya Kovalenko •
Wolfgang Bez • Erich Focht • Hiroaki Kobayashi
Editors

Sustained Simulation Performance 2018 and 2019

Proceedings of the Joint Workshops
on Sustained Simulation Performance,
University of Stuttgart (HLRS) and Tohoku
University, 2018 and 2019

 Springer

Editors
Michael M. Resch
High Performance Computing Center
(HLRS)
University of Stuttgart
Stuttgart, Germany

Yevgeniya Kovalenko
High Performance Computing
University of Stuttgart
Stuttgart, Germany

Wolfgang Bez
Europe GmbH
NEC High Performance Computing
Düsseldorf, Germany

Erich Focht
Europe GmbH
NEC High Performance Computing
Stuttgart, Germany

Hiroaki Kobayashi
Cyberscience Center
Tohoku University
Sendai, Japan

ISBN 978-3-030-39183-6   ISBN 978-3-030-39181-2   (eBook)
https://doi.org/10.1007/978-3-030-39181-2

Mathematics Subject Classification (2010): 65-XX, 65Exx, 65Fxx, 65Kxx, 68-XX, 68Mxx, 68Uxx, 68Wxx, 70-XX, 70Fxx, 70Gxx, 76-XX, 76Fxx, 76Mxx, 92-XX, 92Cxx

This Springer imprint is published by the registered company Springer Nature Switzerland AG.
The registered company address is: Gewerbestrasse 11, 6330 Cham, Switzerland

Preface

Sustained simulation performance is becoming an ever more important issue in High Performance Computing (HPC). Hardware is moving towards the Exaflop, and we will see such systems in the near future in China, Europe, Japan, and the USA. However, sustained performance is lagging behind substantially. Experts are worried that the level of sustained performance will stay as low as 1% of peak performance for typical applications.

The workshop series on sustained simulation performance has set out 15 years ago to tackle this problem. The papers presented here are hence looking into a variety of issues that have an impact on sustained simulation performance in HPC.

The starting point for any such investigation is hardware architecture. The key problem of modern HPC systems is the lack of speed in communication mainly for the main memory. The currently only vector architecture which has the potential to overcome this problem in HPC is the NEC Aurora. Several of the articles in this volume refer to this architecture and its potential for HPC simulation. Based on an excellent architecture, basic software plays a vital role. This includes a variety of topics like operating systems, compilers, schedulers, IO-systems, and programming models. Hardware and software are important for sustained performance, but in the end it is mathematical algorithms that have to be implemented and hence finally decide on how well hardware and software are used. In the coming decades, the optimization of mathematical algorithms might replace Moore's Law as the main driving force in sustained performance on HPC systems.

The contributions in this volume show that the number of problems in sustained simulation performance is high. Some solutions can be seen but for many problems we still have to invest a lot of research. However, if HPC in general and Exaflops systems in particular want to be successful in the coming decades the focus of attention will have to shift from hardware to software and algorithms and funding will have to go into these fields in order to make sure HPC systems do not become heroes with feet of clay. This book aims to make a contribution not only to make readers aware of the problem but also to put some potential solutions on the table.

Stuttgart, Germany   Michael M. Resch
September 2019

Contents

Part I
Future HPC Challenges

R&D of a Quantum-Annealing Assisted Next Generation HPC Infrastructure and Its Killer Applications

Hiroaki Kobayashi

Abstract As the silicon technology driven by so called Moore's law is facing the physical limitation, we are now moving to the post-Moore's era in the design of the next generation high-performance computing infrastructures. Under such a situation, Quantum Annealing is expected to be one of emerging information processing technologies in the Post-Moore's era, which is especially work well for combinatorial optimization problems. In this article, we present our on-going project entitled, "R&D of a Quantum-Annealing Assisted Next Generation HPC Infrastructure and its Applications," which aims to integrate the quantum annealing information processing into a conventional HPC system as an accelerator for combinatorial optimization problems. We also show the designs of target applications that integrate computational science and data science approaches to be installed on the underlying infrastructure, which are expected to play a key role in the realization of the smart city (also named Society 5.0 in Japan).

1 Introduction

In the last several decades, thanks to the silicon technology development so called Moore's Law, computer performance has been improved exponentially. However, as the physical limitation in the silicon fabrications is approaching, we are facing the end of Moore's Law. Under such a circumstance, post-Moore's information processing technologies such as Quantum computing, Brain-Inspired computing etc. is drawing much attention as emerging ones to make a breakthrough in computing. In such a trend, quantum annealing is considered one of promising information processing mechanisms in the Post-Moore's era, because it is commercially realized and available right now, although solvable problems are still small. The quantum annealing is a metaheuristic for finding the global minimum

H. Kobayashi (✉)
Tohoku University, Sendai, Japan
e-mail: koba@tohoku.ac.jp

© Springer Nature Switzerland AG 2020
M. M. Resch et al. (eds.), *Sustained Simulation Performance 2018 and 2019*,
https://doi.org/10.1007/978-3-030-39181-2_1

of a given objective function over a given set of candidate solutions (candidate states), by a process using quantum fluctuations. The D-Wave system, a Canadian startup company, has developed the first transverse magnetic field type quantum annealing Chip and System in the world. By using the commercial systems, many research teams of Google, NASA, Volkswagen, Lockheed, Tohoku University etc. get involved in R&D of emerging applications using quantum annealing especially that need solving combinatorial problems, and successfully show its possibility that could outperform classical HPC systems toward the post-Moore's era.

Quantum-annealing shows a great potential, however, it is not a universal solution to the general-purpose high performance computing. Therefore, even though the performance of quantum-annealing machines reaches the practical level in a couple of the next decades, it still needs classical high performance computing platforms to satisfy a wide variety of computing demands from the practical general-purpose applications. Thus, toward the Post-Moore's era, we are conducting a new research project for the realization of a quantum-computing and classical computing hybrid infrastructure just like a hybrid car with an electrical engine and combustion engine. In this project, we try to realize a transparent access to hybrid computing engines based on the demands from applications, while so called Moore's-type traditional high-performance computing engines are enhanced in a domain-specific fashion such as computing-memory performances balanced architectures, e.g., vector-engines, for memory-intensive applications, dense-computing engines, e.g. GPU, for computation-intensive applications, and de-facto standard computing engines, e.g., X86, for controlling these underlying heterogeneous computing engines including the quantum annealing hardware and providing standard programming environments to install commonly-used open-source applications.

At the same time, we are also interested in co-design and co-development of emerging applications with the quantum-annealing and classical computing hybrid infrastructure. In particular, such applications are expected to be essential for the realization of the Smart City, which is also named Society 5.0 in Japan. According to the 5th Science and Technology Basic Plan 2016–2020 by Japanese Government, the Society 5.0 is defined as a human-centered society that balances economic advancement with the resolution of social problems by a system that highly integrates cyberspace and physical space. To realize the Society 5.0, we have to deploy a cyber-physical system into the society, which realizes a close interaction and convergence between the physical space (real society) and the cyber space (virtual world on a computer) as a social infrastructure. The recent remarkable improvement of AI and ML (machine learning) can exploits higher-order information from the large amount of data collected from the real world through the IoT technology in a practical time, however, the simulation also plays an important role in generating the effective data for AI-ML, because high-quality data could be generated by the high performance simulation with several scenarios, which are not available from the real data due to danger, disruption, and/or serious fatality to our life and systems. On the other hand, to improve the simulation significantly, it needs AI-ML. In the AI-ML steering simulation that we are investigate in this project, AI-ML analyzes the simulation results, optimizes the simulation model, and then steers the simulation effectively. This AI-ML steering simulation definitely become a core technique to realize the cyber-physical system. Therefore, the next generation applications should also be designed based on the integration of traditional computational science approach and AI-based data science approach.

In this article, we present our on-going project entitled, *"R&D of a Quantum-Annealing Assisted Next Generation HPC Infrastructure and its Applications,"* which aims to integrate the quantum annealing information processing mechanism into a conventional HPC system as an accelerator for combinatorial optimization problems. Section 2 describes our system design concept of a new computing infrastructure toward the Post-Moore's era by the integration of classical HPC engines and a quantum-annealing engine. In Sect. 3, we present target applications that would be solutions to the realization of the Society 5.0. Here, we define a workflow for the integrated applications of simulation and AI-ML approaches. Based on the workflow, we discuss two applications: a real-time Tsunami Inundation Damage Forecasting and Optimal Evacuation Planning from Tsunami inundation, and a Digital Twin of a Turbine for power-generator systems. Section 4 summarizes the article.

2 Quantum-Annealing Assisted Next Generation HPC Infrastructure

Figure 1 show a stack representation of the target infrastructure to be developed in this project. At the lowest layer, hardware platforms are configured, and we place the D-wave machine as a quantum-annealing engine and NEC SX-Aurora TSUBASA as classical computing engines.

NEC SX-Aurora TSUBASA is the latest vector system and consists of an X86 engine and a vector engine. The vector engine is composed of a vector processor and a memory subsystem. On the vector processor, eight high performance cores are integrated, and are connected to the memory subsystem at a 1.22 TB/s.

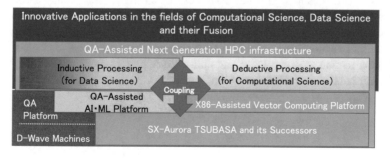

Fig. 1 A stack representation of the target infrastructure

Each core provides 307.2 Gflop/s for double-precision (DP) and 614.4 Gflop/s for single-precision (SP) floating-point calculations. As a result, a VE processor with eight cores provides up to 2.45 Tflop/s (DP) and 4.91 Tflop/s (SP) floating-point performances. This high vector processing capability supported by a high memory bandwidth is expected to realize high-sustained performance in a wide variety of science and engineering applications, especially memory-intensive applications. At the same time, an X86 engine named Vector Host, which is equipped with a Xeon processor, is attached to the vector engine. As the vector host provides the standard LINUX programming environment and OS functions, the LINUX system calls are automatically offloaded from the vector engine to the vector host in program execution as needed. As a result, SX-Aurora TSUBASA have a great potential of high performance vector processing capability, while providing a standard programming environment. At the same layer, we introduce a D-Wave 2000Q as a quantum annealing engine. Program kernels for solving combinatorial problems could be offloaded to this engine through the programming interfaces implemented at the upper level of this layer.

On the hardware engines layer, we construct two fundamental computing environments: a deductive computing environment and an inductive computing environment. The deductive computing environment is prepared to accelerate conventional simulations in the fields of computational science and engineering, and is highly-optimized for exploiting the potential of SX-Aurora TSUBASA. In this environments, effective vector computing-scalar computing hybrid is realized by heterogeneous computing of vector-engines and X86 engines (vector host). We also construct the inductive computing environment for AI-based data science applications at the same level. The inductive computing environment basically consists of application interfaces to the quantum-annealing engine for combinatorial problems and several standard AI-ML platforms such as Tensorflow and SPARK, where the latter ones are highly-optimized for SX-Aurora TSUBASA. Moreover, as it is reported that the quantum-annealing works well in Boltzmann learning and clustering, we try to bring its potential to the AI-ML environment to provide the best choice to users based on their demands in the program development. In this project, we will intensively evaluate performances of quantum-annealing and classical computing as an accelerator for combinatorial problems and AI-ML applications, and figure out their best mix through the R&D of inductive computing and deductive computing integrated applications. Finally, over the these environments, we construct a transparent interface to access these environments in a unified fashion, and build two emerging applications to be presented in the following section as examples of simulation-AI-ML integrated applications, which will contribute to the realization of the Society 5.0.

3 Computational Science-Data Science Integrated Applications

Before describing target applications to be implemented on the quantum-annealing-assisted next generation HPC infrastructure, we define workflow models of simulation-AI-ML integrated applications. Figure 2 depicts the workflow that we define for application developments. There are three models: AI-Driven Simulation model, Simulation-Driven AI model, and their integration model. The AI-Driven simulation model shows a workflow in which simulation is steered by AI-based analysis. After the simulation results are obtained, AI analyzes the results and improves the simulation scenario and the simulation model to effectively proceed the simulation, resulting in the minimization of the user interaction during the simulation process. In the Simulation-Driven AI model, the simulation supports AI by preparing supplemental data, named simulation database, which cannot be or hardly obtained from physical systems. For example, the simulation can generate data by using several scenarios of serious situations in natural disaster and malfunctions of machinery, which we rarely observe but give fetal damage to the society and systems if they happen. The last one is the combination of the AI-Driven Simulation and Simulation-Driven AI in a serial fashion. While simulation effectively steered by AI generates the simulation database, AI analyzes the data from the simulation database as well as real physical data obtained by sensor IoTs, in order to exploit higher order information and/or optimal solutions in the simulation-driven AI model. In the following subsections, we describe our target applications based on these workflow models.

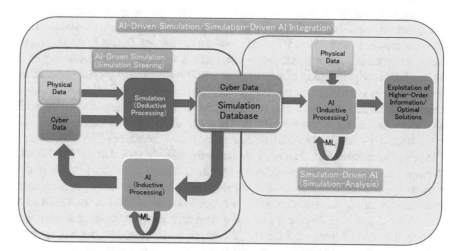

Fig. 2 A workflow of computational science-data science integrated applications

3.1 Real-Time Tsunami Inundation Damage Analysis and Optimal-Evacuation Planning System

The first application we are developing is a Real-Time Tsunami Inundation Damage Forecast and Optimal-Evacuation Planning System. Eight years ago, there was a big earthquake named the 2011 East-Japan Great Earthquake. The earthquake gave us a serious damage, and most of around 20,000 victims were due to the huge Tsunami generated by the earthquake. To prevent and mitigate Tsunami damages caused by big earthquakes, we started a research and development of a real-time Tsunami inundation damage forecasting system in 2014 and have completed the development of the system in 2018. The system can provide the damage estimation of a 6-h Tsunami Inundation in coastal areas around Japan at the 10×10 m mesh resolution in less than 20 min: 10 min for the fault model identification and another 10 min for the Tsunami Inundation simulation on our supercomputer SX-ACE of Tohoku University. The system has already been installed as a part of the disaster analysis system operated by Cabinet Office, Government of Japan in a 24/7 fashion.

In this project, we will be enhancing the performance of the fault modeling by using quantum annealing. The sampling capability of the most likelihood, second, third . . . etc. by quantum fluctuation of quantum annealing could be used to evaluate the certainty/uncertainty of the obtained optimal solution in the fault identification. By reflecting this information to the Tsunami inundation simulation, we can derive the worst case scenario of the Tsunami inundation damage. At the same time, we are also extending the use of the inundation damage forecast information to the evacuation path planning from the inundation. Because the optimal path planning is a kind of combinatorial optimization problems, we try to combine the enforce learning to obtain the first, second and third shortest paths to the safe place with the quantum annealing to obtain the globally optimal path from the combinations of these possible candidates. Of course, we are continuing further optimization of the Tunami code for the latest vector computer SX-Aurora TSUBASA. Figure 3 shows the process mapping of three stages: Fault estimation, Tsunami simulation, and optimal evacuation planning. Purple-colored boxes show the range of hardware engines to be implemented. For example, the first stage "Fault estimation with Quantum-Annealing-enhanced MCMC (Markov Chain Monte Carlo) is on the combination of SX-Aurora TSUBASA and D-wave machine. The second one "Tsunami Inundation Simulation" is implemented solely on SX-Aurora TSUBASA, because the vector processing works well for the Tsunami inundation simulation as a memory-intensive application. The last stage of optimal evacuation planning with quantum annealing is again mapped on the combination of SX-Aurora TSUBASA and D-wave machine. Figure 4 show the workflow of the system. The first and second stages are categorized into the AI-driven simulation, and the last stage is categorized into the simulation-driven AI. As these three stages are coupled in a serial fashion, the total system is categorized into the integration model of the Ai-Driven Simulation model and Simulation-Driven AI model.

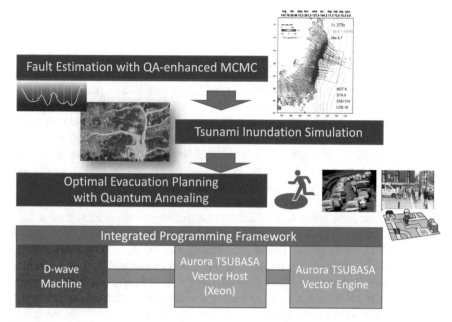

Fig. 3 An organization of a tsunami inundation damage forecasting and optimal evacuation planning system

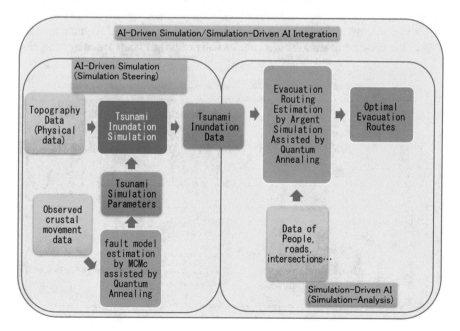

Fig. 4 A workflow of a tsunami inundation damage forecasting and optimal evacuation planning system

3.2 A Digital Twin of a Real-Turbine by Using a Numerical Turbine

The other target application that we are developing in this project is a digital twin of a real turbine. As turbine systems of power plants are a part of important social infrastructures to support our life, their malfunctions and/or failures give a serious damage to the life. Therefore, they need predefined periodical maintenance to keep the systems safe and stable conditions. If we can estimate internal conditions such as aging status of blades from the outside, we can make much more optimal maintenance plans based on the predicted conditions, resulting in a significant cost reduction for the maintenance. To react this demand, we try to realize a virtualized turbine on a supercomputer as a digital twin of the real one. So far, we have successfully realized a virtual turbine named numerical turbine on a SX-ACE supercomputer running at Tohoku University, which can simulates the internal pressure field in the turbine. Therefore, by using several blade conditions such as new one and its gradually aged ones, we can generate internal flows and pressure fields as simulation results of the numerical turbine As the variances in the pressure field represent the degree of aging of blades, and are also expected to be measured as a noise from real turbines, we can estimate the blade conditions by using the numerical turbine. Figure 5 shows a concept of a digital twin of a real turbine by using the numerical turbine. The numerical turbine is used to generate a simulation database with many cases of blade conditions, as well as to design the turbine itself. By combining simulation database and real data obtained by IoT sensors of a real turbine, AI estimates the internal conditions of the real turbine. If AI trained by the simulation database detects the aged blades that needs the maintenance by analyzing

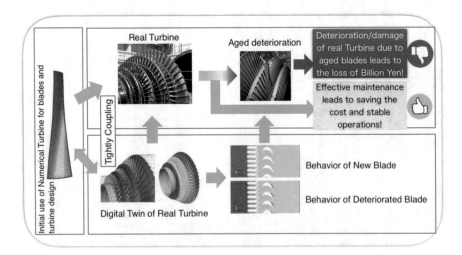

Fig. 5 A realization of a digital twin of a real turbine by using a numerical turbine

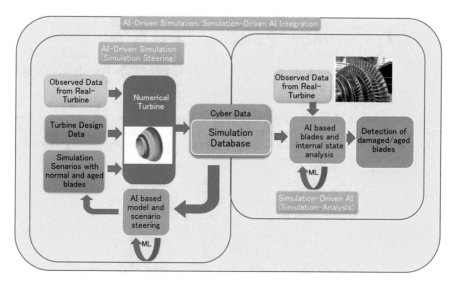

Fig. 6 A workflow of a digital twin of a real turbine by using a numerical turbine

acoustic noises from a real turbine, it could suggest the maintenance before the failure of the turbine system. Figure 6 show a workflow of a digital twin of a real turbine. The first half of the digital twin is classified into the AI-driven simulation model, in which the numerical turbine generates a simulation database with the support of AI to optimize the turbine simulation. The last half is classified into the simulation-driven Ai model, in which AI uses the simulation database for learning the internal states of the turbine, and detects the internal states from IoT-collected data from the real turbine generated by aged blades that needs the maintenance.

4 Summary

This article briefly describes our on-going project for design and development of a future HPC infrastructure, which incorporates quantum-annealing as a post-Moore's information processing mechanism into the classical high performance computing infrastructure. Our design is based on the latest vector supercomputer system SX-Aurora TSUBASA and its successors, but for the specific kernels to handle combinatorial problems, they are offloaded to a quantum annealing machine in a transparent fashion from the user point of view. We are also developing two killer applications for this quantum-annealing assisted HPC infrastructure: a real-time Tsunami inundation damage forecasting and optimal evacuation planning system and a digital twin of a real turbine. These applications are designed based on the combination of simulation and AI-based data processing approaches. This 5-year project started in 2018, and every year we will make progress reports at the workshop on sustained simulation performance to be held twice a year in HLRS and Tohoku University and publications of the post-workshop books as well. After this project completed in 2022, we believe these new generation computing infrastructure and its simulation-AI hybrid applications contribute to the realization of the smart city, so called Society 5.0.

Acknowledgements Many colleagues get involved in this project, and great thanks go to faculty members of the Tohoku-NEC Joint Lab. at Cyberscience Center of Tohoku University. This project is supported by the MEXT Next Generation High-Performance Computing Infrastructures and Applications R&D Program.

Mastering Exascale Challenges
for Engineering Applications

Bastian Koller, Ralf Schneider, Andreas Ruopp, and Dimitris Liparas

Abstract This chapter will present the European approach for establishing Centres of Excellence in High-Performance Computing (HPC) applications, ensuring best synergies between participants in the different European countries. Those Centres are user-centric and thus driven by the needs of the respective community stakeholders.

Within this chapter, the focus will lie on the respective activity for the Engineering community. It will describe what the aims and goals of such a Centre of Excellence are, how it is realized and what challenges need to be addressed to establish a long-term impacting activity in Europe.

1 The Race for Exascale

The race towards Exascale Computing is open and coming to a point in time, where political and technical decisions impact the availability and especially usability of the next level of performance supercomputers. Whilst the US, China and Japan announced their initial plans to make Exascale systems available as soon as possible quite early on, Europe joined this race later, but with a clear and focused strategy. This strategy is embedded and executed by a Joint Undertaking (JU[1])—EuroHPC[2] which was established at the end of 2018 and will be operational until the year 2026. The aim of the Joint Undertaking is to steer European developments in hardware and software and lead to world-wide competitive European Exascale systems, available for (European) users.

---

[1]Definition of a Joint Undertaking—https://eur-lex.europa.eu/summary/glossary/joint_ undertaking.html.

[2]EuroHPC—Leading the way in the European Supercomputing—https://eurohpc-ju.europa.eu/.

B. Koller (✉) · R. Schneider · A. Ruopp · D. Liparas
High Performance Computing Center Stuttgart, Stuttgart, Germany
e-mail: koller@hlrs.de; schneider@hlrs.de; ruopp@hlrs.de; liparas@hlrs.de

© Springer Nature Switzerland AG 2020
M. M. Resch et al. (eds.), *Sustained Simulation Performance 2018 and 2019*,
https://doi.org/10.1007/978-3-030-39181-2_2

Fig. 1 The European HPC ecosystem—applications, hardware and access to infrastructures

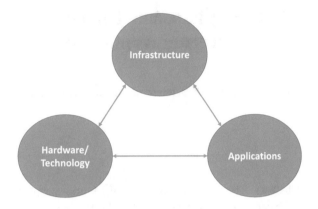

Now, whilst one important pillar for a European HPC Strategy is the hardware development, one of Europe's strengths still lies in the area of applications. Thus, besides the development of European hardware, such as the European Processor,[3] the aim is also to push applications towards Exascale performance, and thus enable European Scientists to access available systems with optimized applications in a best manner.

Figure 1 presents a high-level overview of the EuroHPC strategy, reaching from developments in technology and applications up to the provisioning of access to infrastructures, e.g. as provided nowadays by the Partnership for Advanced Computing in Europe—PRACE.[4] The aim of a European Centre of Excellence for a specific HPC community is thus to identify the potentials of the community codes, to support the communities in the evolution of the codes and finally to showcase what those codes can achieve, once Exascale Computers are available in Europe.

2 European Centers of Excellence

The journey towards the establishment of Centres of Excellence started already in 2015 in Europe. In a first call for proposals, a set of nine community-driven Centres of Excellence was identified. Those targeted the domains of Biomedicine (BioExcel, CompBioMed), Energy (EoCoE), Material Science (MAX, NOMAD, E-CAM), Global Systems Science (CoeGSS), Weather and Climate (EsiWACE), and Performance Optimization (POP).[5]

---

[3]European Processor Initiative—https://www.european-processor-initiative.eu/.

[4]PRACE—http://www.prace-ri.eu/.

[5]Overview of the funded CoEs in 2015—https://www.top500.org/news/Centres-of-excellence-europes-approach-to-ensure-competitiveness-of-hpc-applications/.

Whilst those funded Centres were in average active for 3 years, not all of them were extended in a second call, which was published in 2018. In addition, within this call, topics that had not been addressed so far were identified and addressed. Especially the engineering domain was highlighted as a need, as the European Industry is very strong in that domain and can benefit from a supportive European Activity.

3   Engineering as a Target for Exascale Developments

The European engineering industry consists of 130,000 companies of diverse sizes. Overall, these companies employ over 10.3 million people, with high levels of qualifications and skills. Together they generate an annual output of around EUR 1840 bn and about 1/3 of all exports from the EU. The European engineering industry plays a key role in realising the goal of increasing the industrial production value above 20% GDP by 2020.[6] To achieve this aim and meet the challenges of the fourth wave of industrialisation, it is essential to support European engineering companies in their use of HPC and High-Performance Data Analytics (HPDA), thus increasing the industrial competitiveness of Europe.

The EXCELLERAT project[7] was created to address the requirements of the engineering community with respect to the move towards computing resources with Exascale capabilities. Thus, the aim of the EXCELLERAT Centre of Excellence (CoE) is to boost the competitiveness of European engineering through excellent research that addresses grand challenges of complex applications using cutting-edge HPC technologies and leading to Exascale readiness.

Therefore, the aim of a European Centre of Excellence in Engineering can only be to support the engineering community at a level that no single HPC provider can do and to ensure that the European knowledge in this domain is used synergistically to provide the best possible support.

3.1   The Questions to Be Answered: The Challenges to Be Addressed

In general, EXCELLERAT is designed to become an operative legal entity, acting as a Centre of Excellence for Engineering in Europe (and beyond). Whilst the technical excellence is given by the set-up of the consortium behind this activity, business aspects will also need to be addressed during the implementation and evolution of the Centre.

---

[6]http://www.eesc.europa.eu/m?i=portal.en.ccmi-opinions.34832.

[7]https://www.excellerat.eu/.

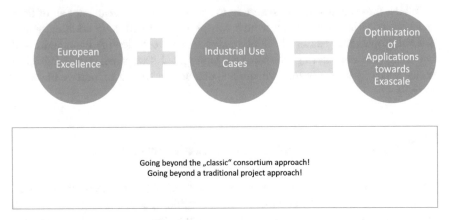

Fig. 2 The EXCELLERAT vision

That implies that the consortium will face many more challenges than merely technological ones. Nonetheless, this subchapter will focus on how EXCELLERAT will tackle technological challenges, whilst the business impacts will be discussed in Sect. 4.2.

Figure 2 presents the approach of EXCELLERAT to tackle the challenges of evolving engineering applications towards the best possible use of Exascale capabilities in Europe. This will enable, amongst others:

- Simplified access to knowledge
- Simplified access to data
- Simplified access to systems
- Clear understanding of benefits from the use of HPC
- Tailored training and education activities

Thus, the Centre aims to act as a single access point to technology and expertise. Academia and industry shall avoid overlapping investments by making use of the knowledge pool of EXCELLERAT. The desire is to give all parties the chance to free up their own resources to drive niche innovations specific to a particular code.

3.2 Addressing the Whole Engineering Life Cycle

The overall approach towards the realization of the Centre of Excellence on Engineering applications is based on the Engineering Work flow, as defined in Fig. 3. EXCELLERAT has chosen the approach of addressing the full engineering work flow, where four different phases and corresponding activities exist. This implies bringing together experts, code owners and stakeholders from various areas and providing expertise on development aspects in interconnected domains. Thus, the EXCELLERAT work force consists of consortium members, bringing together all

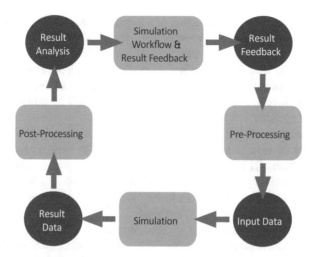

Fig. 3 A high-level view of the Engineering Work flow

the necessary expertise to fulfill the mission.[8] Hence, EXCELLERAT will focus on setting up the necessary mechanisms, which will ensure support to developers and application end users working within every phase of the engineering simulation work flow. These mechanisms will use the tailored services of the Centre and provide links to already existing services from other initiatives and individual HPC Centres. Note that generally speaking, simulations are employed in the product development cycle in engineering to identify design measures and solve issues with the current product configuration, or handle optimization tasks.

4 EXCELLERAT: The European Center of Excellence for Engineering

To implement EXCELLERAT successfully and to future-proof it, we decided to follow an application-driven approach to ensure the "real-world" applicability of the Centre's outputs and services. As the driving applications, six reference applications from different engineering domains have been selected (namely Nek5000, Alya,[9] AVBP,[10] Fluidity,[11] FEniCS and Coda (aka FLUCS[12])).

---

[8]https://www.excellerat.eu/wp/about/partners/.

[9]https://www.bsc.es/research-and-development/software-and-apps/software-list/alya.

[10]http://www.cerfacs.fr/avbp7x/.

[11]http://fluidityproject.github.io/.

[12]https://www.excellerat.eu/wp/engineering-applications/codes/.

One major criterion for the selection of the applications was their already proven efficient deployment on current Petascale HPC-Systems. Along with each core code at least one use-case of an end user of this application has been selected, each of which, together with the respective reference application, creates requirements in terms of technical (e.g. porting to new architectures) and/or intellectual (e.g. algorithmic improvements) developments to unfold its full usability and impact on Exascale HPC systems.

These requirements will be evaluated from the algorithmic, as well as the hardware point of view to initiate an iterative co-design process tailored to engineering applications. This process will lead to improved implementations and new methodologies in terms of software and hardware that will enable the engineering community to successfully master the technical challenge of harvesting a maximum of efficiency of novel Exascale HPC-systems.

With the main overarching goal of strengthening the European excellence and technological leadership, we apply efficient mechanisms for software development and validation, quality assurance and support in the project. All functionalities will be developed or integrated incrementally, continuously evolving the EXCELLERAT Centre of Excellence and service portfolio and being immediately available for the reference applications and further uptake by other engineering applications.

As EXCELLERAT will not only provide technical solutions, but a full Centre of Excellence, a variety of other, not necessarily plain technical activities is provided as (so-called "side services"), to complement the focus on applications and to get fully functional as a Centre.

Finally, a path towards sustainability of the Centre and its services is of utmost importance and the path towards a full understanding on what sustainability means is part of the overall activity.

4.1 The Questions to Be Answered: The Challenges to Be Addressed

As mentioned before, the work within EXCELLERAT will not be successful, if the Centre focuses solely on sub-parts of the engineering application life cycle. Thus, a clear identification of the challenges in the life cycle and technological parts of the applications is needed and subsequently needs to be addressed to ensure Exascale capabilities of codes.

Furthermore, the overarching goal shall not only be to support proper execution of applications on future Exascale systems, but also to be able to deal with so-called side services, which are intended to support users, e.g. with secured data transfer capabilities or visualization of results. Figure 4 shows some of the activities, which we see as being challenged by the move towards new and more complex supercomputers. As this is just a subset of the so-far identified challenges, we expect this to be a living list that will be updated and extended on the basis of further

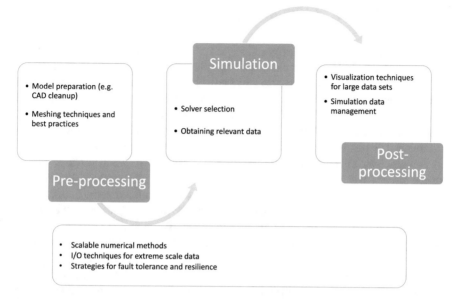

Fig. 4 A deeper view on the Engineering Work flow

expertise gained when working with the aforementioned six reference applications. EXCELLERAT focuses its developments and thus the evolution of the reference applications on a set of twelve use cases, which represent challenges within the applications. Thus, it is ensured that whatever is produced, is steered by and made for the enhancements of the reference applications.

For example, innovations are foreseen on topics in the areas of adjoint methods, meshing algorithms, multilevel discretization strategies, data reduction algorithms, matrix free solution methods, node and system level performance engineering. Furthermore, the porting to new architectures and the usage of new programming models and paradigms will be elaborated. In addition to pure optimization, focus will also be given to enhance the capabilities of the core codes, so that new simulations can be enabled.

EXCELLERAT will introduce advanced in-situ data analytics methodologies into selected codes, in particular with a view on the advanced statistical, mode and comparative analysis of simulation runs approaching Exascale. The analytics service will be based on an innovative data management approach, including reduced features that capture the essential behavior, which are stored and later employed for further analysis. In addition, the integration with data visualization allows for the complete (visual) exploration of a computed, data-driven design space.

For co-design, existing relationships will be built on and expanded across the EXCELLERAT Centre, and new ones developed with HPC stakeholders in order to ensure that the co-design stack is considered in its entirety, from the circuits on the CPU to the high-level engineering application.

4.2   The EXCELLERAT Strategy

To summarize, the global strategy (cf. Fig. 5) of EXCELLERAT focuses on a user-centred view, namely, in our case, the six different reference applications. Based on those, developments related to diverse technical topics, such as Node-Level Performance Optimization, System-Level Performance Optimization, Advanced Meshing or Data Handling and support activities in the areas of Co-Design, Visualization, Data Analytics or Data Management are performed, elaborated and established as part of the service portfolio of the Center of Excellence.

Nonetheless, the developments and the final strategy on which services will be really offered within the Centre will also need to take into account the business aspects, to ensure sustainability of the Centre beyond the point that the funding from the European Commission ends. This includes activities to ensure that the market for the service offerings of the Centre is understood, that a proper business model is developed and that a sustainable operation is set up. This will allow to implement the necessary structures for a sustainable operation: the management, the development of the products/offerings, the continuous operation and the communication activities that are necessary to attract potential users and convince them to participate in and make use of the offerings of the Centre.

Specifically, the development activities for services, as well as products will be advised so that they correspond to the market needs. Finally, the legal frame of the operation of EXCELLERAT will be developed and set up in the best possible manner, to allow the development of a stable cooperation of all partners involved and to address important aspects, such as protection of the intellectual properties of the active partners.

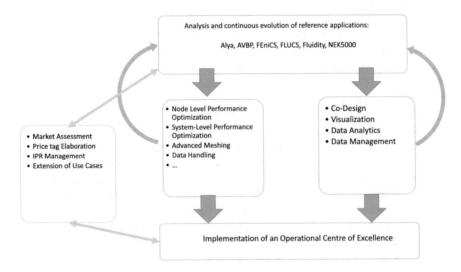

Fig. 5 The EXCELLERAT strategy

4.3 Increasing Impact: Introducing the Concept of Interest Groups

As described in Sect. 3, sustainability is one of the implementation goals of EXCELLERAT. Whilst one part is the development of the business model, the other part is to understand the market, especially with regard to potential end-users, but also potential new partners for the Centre of Excellence. Thus, it is of utmost importance to increase the impact of the Centre by reaching out to external entities already in the EXCELLERAT set-up phase.

For that purpose, the concept of Interest Groups was developed, which is the mechanism to integrate externals into the EXCELLERAT evolution and thus completes the potentially focused view of the consortium members by enlarged knowledge and advise (cf. Fig. 6). The Interest Groups will be set up at the beginning of the project and consist of selected representatives from external entities (external to the consortium) which are integrated into the EXCELLERAT evolution process. As members of these groups, they will get first-hand access to the project results (partially under NDA) and will be able to deliver quick feedback from their own perspective.

The participation in the Interest Groups is on an invitation only basis to ensure that focused discussions are possible. In terms of entities, four interest groups are foreseen, which can provide the viewpoints of the different roles of the Centre's value chain: Code Developers/ISVs, Industrial End Users, Scientific Experts and Technology Providers.

In a long term, the Interest Groups shall also act as an exchange forum to identify common problems and thus, build community support to develop ideas for potential satellite activities.

Fig. 6 The EXCELLERAT Interest Group concept

5 Conclusions

This book chapter has reported about the ongoing activity of establishing a European Centre of Excellence in Engineering—EXCELLERAT. At a glance, the major ideas behind the envisaged activity have been sketched and the goals and strategy of the work to be performed have been outlined. In addition, it is clear that with respect to long-term sustainability, one will only be able to succeed, when not only technological issues are addressed, but also business aspects are taken into account. The same level of importance will have an early outreach to potential future customers of the Centre's services, which will be manifested within the implementation of the Interest Group concept.

Acknowledgements This work has been supported by the EXCELLERAT project which has received funding from the European Union's Horizon 2020 research and innovation programme under grant agreement No 823691.

Some Thoughts on Processor and HPC Hardware Technology

Thomas Bönisch

Abstract Having the end of Moore's law and the end of Dennard scaling in mind, this paper looks into current developments in the hardware market. Based on this, it shows the impact of this development on sustained application performance and gives some hints, where future hardware architectural improvements might help engineering applications for a better sustained performance.

1 Motivation

For some years now, we can recognize, that Moore's law [1, 2] is coming to its end. This law, saying that every 24 months the number of transistors per area is doubling, was valid for nearly half a century and it has guaranteed an increase in processor performance as the frequency was increasing as well. This is nicely visible in the development of the top500 [3]. In Fig. 1 one can see the flattening in the performance development in the last years compared to the decades before. To show this, looking at the system ranked 500 is even more valid than the number 1 system, as the number one system is also influenced by political decisions. A second important law was the Dennard Scaling [4], saying that the power usage is linear with the used chip area. This law is not working any more, too. And this imposes an even more severe issue on the hardware developers and the computing centers, as this directly influences the costs for power supply and cooling. TSMC, one of the foundry operators, announced [5], that for the upcoming 5 nm process a power degradation by 20% is estimated compared to the current 7 nm process, by not changing the

T. Bönisch (✉)
High Performance Computing Center Stuttgart (HLRS), Stuttgart, Germany
e-mail: boenisch@hlrs.de

M. M. Resch et al. (eds.), *Sustained Simulation Performance 2018 and 2019*,
https://doi.org/10.1007/978-3-030-39181-2_3

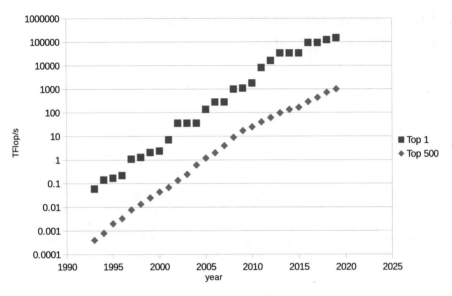

Fig. 1 The performance development in the Top500 list [3]

chip frequency nor the chip complexity. At the same time, they can put about 80% more transistors into the same area. Calculating this for a future processor in 5 nm technology, it means about 80% more performance compared to a 7 nm processor with the same chip size. But at the same time, power and cooling demand increase by a factor of approx. 1.44. Thinking this further to a 3.5 nm process, if this will happen, we can make a best case estimate for performance and power consumption. This would mean a performance increase by about a factor of 3.2. At the same time power requirement would go up by a factor of 2.1. Seeing this, we can recognize, that performance improvements just by improving the chip manufacturing process are quite limited for the future. However, the chip manufacturing process is not the only parameter for a good sustained performance of the running applications on a computer or a supercomputer. There are several more factors influencing sustained performance. Of course one important factor is the style of programming which might easily lead to bad application performance if not done well. On the other side, there are several hardware developments which have made getting a good application performance quite hard in the last years. So I propose hardware support for a good application performance within this paper. In Sect. 3, I will sum up some recent developments. In this chapter, I will discuss the performance issues with those developments and I will make some proposals about what should be improved for higher sustained performance. Section 4 concludes this paper.

2 Recent Developments

2.1 Server CPUs

In the last years, the server CPU market has been dominated by one company. Performance improvements in processors have been obtained by three major developments:

- increasing the number of cores
- increasing the vector length to eight double values
- increasing the number of memory channels

Of course there have been several architectural improvements in addition, but they have not been as performance relevant as the mentioned three. Clock frequency has been mostly stable, so there has been no performance improvement by just clocking up the CPU. Moreover, in several cases the clock frequency went down compared to predecessors to be able to run the higher number of cores. Depending on the hardware vendor, there have been slightly different design decisions out of the three major development directions. On one hand, we see processors with a core count of up to 64 but a vector length of four double precision values and providing eight memory channels of the latest speed. On the other hand, we have processors with up to 28 cores and a vector length of eight and six memory channels. In addition, there are also modules, where two of those chips have been bounded together.

2.2 Accelerators

The accelerator market has been developing fast in the recent years. But this development in the area of GPGPUs headed mainly in the direction of machine learning and deep learning. The key features here are tensor units with half and single precision accuracy which cannot be used that much in current engineering applications. Nevertheless, there are quite a number of double precision units available which lead to peak performance values that are unreached by general purpose server CPUs.

In addition, there is a vector accelerator available. This card works mainly like an independent CPU which is taking over the compute intensive applications from the host. In addition, there is an offload mode available.

2.3 Main Memory

On the memory side, high bandwidth memory (HBM) has been developed, which is now available in its second generation. This memory provides a quite high memory bandwidth with up to 256 GB/s per stack. But, its capacity is limited by the stack size of 8 GB. In addition, there is no latency improvement compared to standard DDR4 memory, the latency is even a bit higher. In addition, the cost of HBM is not negligible.

2.4 NVRAM

In the moment, we see the advent of non-volatile memory (NVRAM). This memory can be used as byte addressable memory or as block addressable storage. Using it in memory style allows for a quite high memory layout per node as one module already provides up to 512 GB capacity. However, memory bandwidth is less and memory latency is much higher compared to standard DDR4 RAM.

2.5 Network

In the field of high performance networking, network bandwidth went up in the last years. Meanwhile we see networks with up to 200 Gbit data rate per link. In addition, several topologies are supported. However, network latency has not been increasing.

3 Prospects and Issues

What does those developments now mean for sustained performance. Here we mainly have a look to engineering applications as they are the main focus of HLRS.

3.1 Server CPUs

Actually, the current vector facilities deployed in micro processors have their limits. First, the vector length is rather short. Second, to fully make use of the units, the data has to be vector register length aligned. This means, the beginning elements of a vector which does not start at a proper memory location will be handled differently and with less performance. Same for the last elements of a vector as long as they do not fit completely into the vector register. An additional problem is, that the current processors get their performance out of a high core count in combination with the vector units. Typically, applications try to benefit from the high core count by shared memory parallelization. This now means, that with vectorization and shared memory parallelization the parallelization of the same loop structures are addressed.

Depending on the application and the used numerical scheme for simulation, there is potentially not enough meat in the loops to fully vectorize the loop and to parallelize the loop structure to more than 100 cores. In these both cases properly setup vector units with flexibility concerning the vector access and the data dependencies in a loop could be of help. Another issue with the high core count working in a shared memory parallel fashion in the codes is the effort for synchronization and global structures like global sums, critical sections or loops. As these structures are often unavoidable, some hardware support at least at a socket level would be desirable. Probably, there is the possibility to provide something like a shared register or a directly addressable fast piece of memory for that on the die which can quickly be accessed by all cores of a socket more or less directly.

3.2 Accelerators

Up to now, GPGPUs have not played a big role in HPC in the engineering field. The SIMD style of programming is not so helpful, the memory transfer speed to the host is limited by the bus and programming in CUDA for best performance is complicated e.g. with the large CFD applications containing millions of code lines. This might change, if the GPGPU gets closer to the processor, i.e. it can directly access the host memory with full speed. Then, it is possible to offload parts of an application more efficiently to the accelerator. This would allow for using the accelerator only in selected parts of the application where there is benefit from its usage.

The new vector accelerator Aurora-Tsubasa from NEC is also a step in the right direction. However, today's engineering applications are not so easy to vectorize as in former times. This is mainly due to the fact that those applications do quite some work in organizing the mesh hierarchy and the overlay meshes. In addition, the resulting load imbalances have to be handled. Potentially, bringing the vector accelerator closer to the CPU would help here, too.

3.3 Memory

With the current and the potential future memory structure, there are quite some issues, too. First problem is indirect memory access. This kind of memory access happens often in today's engineering applications, e.g. when using unstructured meshes. Another source of indirect memory access is the usage of adaptive mesh refinement or overlay meshes. All those are state of the art programming techniques. However, memory latency is really an issue for this kind of system usage. First we hoped for a better memory latency with the further development of the HBM memory. But meanwhile, we have understood, that memory latency is not a development focus for HBM3 and later versions. So, we cannot expect less memory

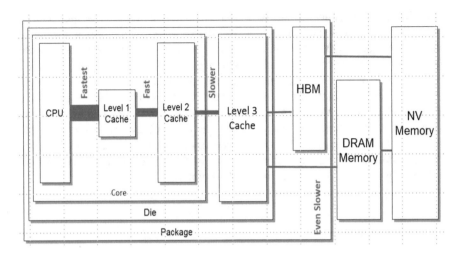

Fig. 2 Assumed memory hierarchy on a future HPC system node

latency with this technology. Potentially, DDR5 memory might help here a bit. But this depends on the used RAM and so it is not clear, whether there will be less memory latency in the future. Nevertheless, there might be an additional solution for this problem. Even though the memory access is indirect and in a sense unstructured, the access pattern in our cases do not change much in between different access epochs like a time iteration. Therefore learning the access pattern might help here. Probably, there is some hardware in the future which can take advantage from this.

A second issue in the memory field is the future memory hierarchy. Figure 2 shows how the future memory layout of a node might look like. Unfortunately, all three types of memory are required as they have different characteristics:

- HBM: high bandwidth but limited capacity and quite some latency
- regular DDR4/5 memory: less bandwidth than HBM, but better latency and more capacity
- NVRAM: besides persistency, it offers high capacity compared to HBM and DDR memory, but providing much longer latency and much less bandwidth.

The question is, how this new hierarchy will be made accessible for the user:

- by compiler directives for placement (only)
- by some hardware support like the caches are managed today
- by intelligent placement algorithms powered by machine learning

All those solution options will definitely need more elaboration and development.

3.4 Network

In the field of HPC networks, we have one major issue. In the last years there has been no latency improvement at all. Bandwidth went up, but on the latency side there was no further reduction available. On the other side, several engineering applications send a high amount of rather small messages especially if the application is still an MPI-only implementation. Meanwhile, we have seen applications spending 70% of their time within MPI. So, a further reduction of the HPC network latency would be desirable. One solution for that might be to bring the HBA/HCA closer to the CPU. This would mean an integration of the network adapter on the interposer or even directly on the die as some Systems on a chip (SoC) do it already for Ethernet. Such a solution would help a larger class of applications a lot.

3.5 I/O

The I/O field is unfortunately often ignored by the programmers and users of scientific applications. When making a ranking of topics where programmers are working to optimize their HPC applications, there is first scalability. If they are interested, they spend additional time in single node and single core performance. I/O performance comes last and it is mainly ignored. In the best case applications are using an I/O library like HDF5 and programmers think that this is the solution for their issues. Unfortunately, this does not per se help for I/O performance especially when running the application on a huge number of nodes. We have seen applications using 50% of their run time in I/O after having done quite some optimization already. What might help here in the future is the use of NVRAM when reading and writing data from the application. The data transfer from NVRAM to disk or another storage with long term persistency is then done before application start and after the application has ended. To make this happen quite some effort is required. It has to be clarified, where the NVRAM is best located in a system: at the nodes or in a central part of the system or probably distributed over a number of NVRAM nodes/NVRAM islands. A further decision is the connection of the NVRAM to the processors: by usual DDR memory channels or by a standard interface like OpenCAPI or CCIX or by a special kind of interconnect. All those solutions have their pros and cons which need to be further investigated. A third requirement is the availability of an appropriate software which supports this I/O usage model.

4 Conclusions

As we have seen, there are many approaches to taggle the performance issues of today's engineering applications by improving the hardware, mainly in the hardware architectural field. Doing so, the end of Moore's law might even be an advantage.

As the processor performance is not automatically doubling every 2 years because of technical progress in the fabrication process of the processor, there is potentially much more time to develop and test architectural improvements. In addition, and this is most probably much more important, a developed processor architecture might sell not only about 1.5 years but much longer. This means the number of potentially salable processors of a type is becoming higher. This opens the door for architectural improvements which help only a class of users or applications and not the whole community.

With the changes in hardware, a second point is the availability of appropriate software in all required fields, from the operating system level over the middleware up to compilers and tools. This is required to fully elaborate the hardware potential. How, for example, can an improved processor unit or an improved instruction set be of use, if a compiler is not able to make use of it? Or how can an improved network help, if the middleware is not able to handle it properly? This means, the solution of our problem is not only hardware improvement, but, it also requires a co-design approach over all hardware and software layers. Such, it could be decided together which issue to handle where and in what way, finding the best solution for a system. This is most probably the hardest problem as it requires a lot of communication and understanding. This means not only technical understanding but speaking the same language, using the same terms for the same thing and the will to work together. Nevertheless, this is the only way to make progress to a high absolute performance of future (scientific) applications. And this is a requirement for further scientific progress in the different disciplines which depend on simulation technology.

References

1. Moore, G.: Cramming more components onto integrated circuits. Electronics **38**(8), 114 (1965)
2. Takahashi, D.: Forty years of Moore's law. In: Seattle Times. San Jose (2005). https://www.seattletimes.com/business/forty-years-of-moores-law/. Accessed 24 Sep 2019
3. Meuer, H.W., Strohmaier, E., Dongarra, J., Simon, H., Meuer, M.: The Top500 list. https://www.top500.org/. Accessed 24 Sep 2019
4. Dennard, R., Gaensslen, F., Yu, H.-N., Rideout, L., Bassous, E., LeBlanc, A.: Design of ion-implanted MOSFET's with very small physical dimensions. Proc. IEEE **87**(4), 668–678 (1999). Reprint from IEEE J. Solid State Circuits **9**(5) (1974)
5. Shilov, A.: TSMC's 5nm EUV making progress: PDK, DRM, EDA tools, 3rd party IP ready. Anandtech (2019). https://www.anandtech.com/show/14175/tsmcs-5nm-euv-process-technology-pdk-drm-eda-tools-3rd-party-ip-ready. Accessed 24 Sep 2019

Part II
Performance Analysis and Optimization on Modern HPC Systems

Overall Project View at HLRS as the Basis for Optimizing Applications

Björn Dick, Thomas Beisel, and Manuela Wossough

Abstract HLRS, the High Performance Computing Center in Stuttgart, runs one of the most powerful HPC systems in Europe. As HPC systems are expensive resources, one of the main objectives of HLRS is to maximize scientific profit and progress from their usage. In order to achieve this, HLRS wants to ensure an *efficient* utilization of the HPC resources and advises its users in order to do so. In addition, HLRS committed to improve its sustainability. For data centers, this can be done by optimizing the power consumption. Since performance optimization shortens computing time, it is in a sense transferable to saving energy. Also, in future computer architectures, performance optimization will no longer be achieved by new hardware. Performance bottleneck detection and performance optimization of codes and applications therefore is one of the key challenges in HPC.

1 Lifecycle of Projects Using HPC Resources at HLRS

At HLRS, computing time is used by industrial research projects as well as EU research projects and projects in context with solution centers. However, the major share of computing time is used by academic research projects. All academic research projects go through a well-defined lifecycle which consists of

- project proposal submission and evaluation
- production phase of the computing time project including training and support for users
- project evaluation
- publishing project results

Submitted *project proposals* contain basic information on technical details like requested resources, scientific details like the description of the scientific problem

B. Dick (✉) · T. Beisel · M. Wossough
High Performance Computing Center Stuttgart, Stuttgart, Germany
e-mail: dick@hlrs.de; beisel@hlrs.de; wossough@hlrs.de

© Springer Nature Switzerland AG 2020
M. M. Resch et al. (eds.), *Sustained Simulation Performance 2018 and 2019*,
https://doi.org/10.1007/978-3-030-39181-2_4

and the approach towards its solution as well as information about previous work in this field. Also information on preliminary studies which show the scaling behavior as well as serial behavior of the program codes under production conditions is highly required. This includes typical parameter sets and problem sizes of the planned project. Furthermore, a detailed description of the I/O behavior, e.g. number and size of files generated during typical runs, the used I/O strategy as well as an estimate of overall storage requirements is mandatory. The latter is necessary since I/O is getting more and more important these days. The applicants must have the know-how which is necessary for efficient usage of high-end computing systems, proven e.g. by presenting work done on smaller computing systems, scaling studies etc.

After submission, *project proposals* are evaluated with respect to different aspects. Apart from assessing scientific goals and challenges, also technical aspects of the simulation code (like parallelization strategy, scalability and efficiency) as well as data management (e.g. I/O and data transfer strategy) are considered while deciding on computing time to be granted.

Proposal evaluation therefore figures out projects which meet the basic requirements to run efficiently on the HPC resources.

Within the *production phase*, projects are reviewed by assessing their annual reports. Outstanding project results are presented in the annual *HLRS Results and Review Workshop*. Furthermore, there is a project evaluation after the end of the project. Both reviews mainly focus on scientific results in the respective application domain (computational fluid dynamics, material science, chemistry, etc.). However, in the future HLRS will provide key performance indicators for the project evaluation to enable also aspects of efficient usage of the HPC resources like code performance, I/O performance and filesystem usage to be assessed. Moreover, HLRS encourages its users to also include aspects related to the efficient usage of HPC resources when publishing project results.

2 Current State of Efficiency Optimization Measures

The aim of all the efficiency optimization measures currently implemented at HLRS is to raise the algorithmic *as well as* parallel *as well as* serial efficiency of user codes and their I/O. It's worth noting this since the majority of the HPC community these days focuses on algorithmic and parallel efficiency only.

Furthermore, HLRS is interested in fostering the efficiency of *actual production* jobs, as opposed to versions that have been stripped down for benchmark purposes.

As efficiency optimization is a time-consuming task, one has to decide whom to serve first. HLRS does this based on the remaining unused computing time of the projects. This is due to the fact that *unused* computing time can still be used in an efficient manner in case of efficiency enhancements.

2.1 Training

The first pillar of efficiency optimization measures currently implemented at HLRS is a series of optimization workshops established in 2016. The workshops are held twice a year with a duration of 5 days each. The idea of the workshops is to bring together application as well as hardware expertise in the shape of users and HLRS staff. In order to do so, every participating project is getting fulltime support by an HLRS employee during the workshop. Those tiger teams start with an analysis of the code's runtime behavior. Based on the results, requirements of the user (flexibility, portability, maintainability, etc.) as well as those of HLRS (efficiency) are intensively discussed in order to figure out a balanced solution. We believe this to be essential, since otherwise users will neither accept nor use the optimization to be developed. Afterwards HLRS staff will help the user with implementing the optimization found. This process is repeated again and again in order to address multiple bottlenecks. The HLRS staff involved consists of experts in fields like node-level performance, MPI and I/O, but also experts in computational fluid dynamics, molecular dynamics, quantum chromo dynamics etc. are involved to ease communication of HPC and domain experts. Furthermore, project supervisors are involved because they are aware of the project history (problems already tackled before, potential boundary conditions not mentioned by users, etc.), which might be beneficial in order to develop effective optimization strategies.

Apart from this workshop, further courses and trainings with respect to efficient usage of HPC resources are offered by HLRS.

2.2 Continuous Collaboration with Users

At HLRS, user support is available during the entire production phase. This includes not only basic support regarding system usage (e.g. access to systems, workflow, etc.), but also support to foster efficient execution of the code. In order to do so, the tiger teams mentioned above are continued after the optimization workshops if required.[1]

2.3 Automated Performance Tracking

Collecting performance information on codes is necessary and helpful in order to identify those codes which run inefficient and need improvement.

---

[1]This is possible due to kind funding by the Ministerium für Wissenschaft, Forschung und Kunst Baden-Württemberg via the project SiVeGCS as well as by the European Union via the project HLST (High-Level-Support-Teams).

Performance analysis tools provided by Cray and—with introducing the next generation super computer—HPE are used to gather performance relevant measured values and analyze parallel/serial efficiency and I/O of the relevant application.

On the Cray XC40 system at HLRS, huge amount of performance data is captured using a Cray software called LDMS. This software is able to read node level performance data (e.g. cache misses, memory bandwidth, usage of CPU functional units) as well as energy usage and performance counters of the high speed network. The sampling rate of 5 s and the system size (7700 nodes, 180,000 cores) produce about 1 TB data per day. Users get an extract of their job related performance data written in the HOME directory. To reduce the amount of disk space required to store this performance data, data is kept for 12 weeks only.

Apart from that, a tool called Ludalo has been developed by HLRS to analyze I/O usage on Lustre filesystems. The performance data is collected in a centralized database. In order to do so, the Lustre OST servers send performance data using socket connections to the data collector process which is running on the database server. Afterwards, the latter inserts the measured values into the database. Within the database, also job information collected by the batch system is available. Users and system administrators can access I/O usage at runtime and display those of already finished jobs using a graphical interface (Fig. 1).

SAIO, a Semi-Automated I/O Tuning Framework developed at HLRS, uses a machine learning approach to determine the optimum configuration of Lustre stripe counts, stripe size and MPI collective read and write for a given application. This tool is designed to be portable across multiple HPC platforms and requires little knowledge of parallel I/O optimization. As a wrapper for the MPI-IO library, it is compatible with the parallel HDF5 and parallel NetCDF libraries.

2.4   Hardware Failures

With exascale on the horizon, it's important to handle hardware failures in a way so that overall system performance is not affected. HLRS hence also cares about this. With respect to the current flagship system (Cray XC40, codename "Hazel Hen"), hardware faults will be repaired by Cray onsite staff and all failures will be logged into a Cray-internal DB while HLRS can get a copy of HLRS related incidents. With respect to the cluster systems operated by HLRS, hardware logs are placed in a section of the trouble ticket system. For the next generation supercomputer system, all hardware failures will be logged locally. The correlation of incidents with the accounting data may provide useful information to find applications which produce heavy load on nodes or subsystems and hence trigger failures. This applications (or similar testcases) may be used for stability tests to improve the overall system reliability.

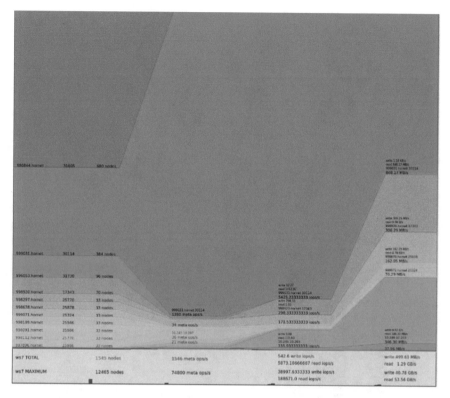

Fig. 1 The Ludalo GUI provides information for the 10 active jobs consuming most I/O resources. The X-axis describes the job size, number of I/O calls as well as read and write operations. The Y-axis displays the relative usage of the resources. A click on the job name will open a window to display the I/O usage of this job over time

3 Future Plans

In the future, HLRS plans to collect *all* performance related data (usage of CPU, memory subsystem, network and I/O as well as job and hardware failure data shown in Fig. 2) in a *unified* database. (In order to do so, an important task will be to implement data reduction due to the high data rate generated by different performance data collection systems. This might be done by a combination of round robin databases and aggregation of smaller unimportant jobs. To be able to get a system wide overview, deletion is not an option.) Such a system improves user support and enables system wide performance analyses as described below.

Improve System Configuration: By means of the *overall* performance data collection in the unified database, it will be possible to identify bottlenecks of the system configuration if a large number of jobs suffers due to a *common* bottleneck.

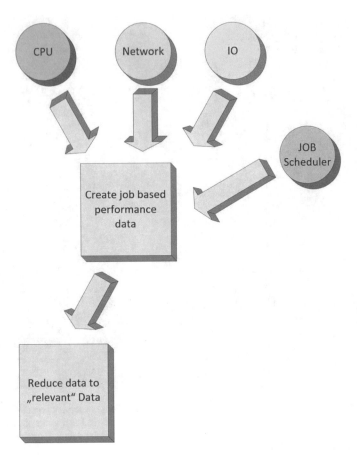

Fig. 2 HLRS plan to collect different performance data sources, reduce the amount of data and create relation to user jobs

Improve System Stability: By correlating job, performance and hardware failure data from the unified database, it will be possible to detect situations where computing load triggers hardware failures (HLRS already faced this in the past), and hence improve system stability. If there are applications which cause extreme load to system resources, the respective users will be asked to provide a simple test case to enable regression tests. By doing so, a test suite will arise over time which allows to check (e.g. at the end of a maintenance) whether all components work as expected and none will fail even under heavy load conditions. Within such a regression test, also performance values of system components can be compared to expected values (cf. below).

Analyze Performance: A major purpose of the unified database will be to analyze the performance of user jobs in order to figure out those to be optimized. By doing so, one can investigate the performance of actual production jobs instead of artificial benchmark cases with non-representative characteristics. In order to do so, HLRS plans to deploy data analytics as well as artificial intelligence/machine learning approaches. Aspired usage scenarios include, but are not limited to:

- Learn performance characteristics of a specific user or code and trigger a notification if a job deviates from those. Such an alarm might inform users about performance degradations introduced by a code patch, but also system administrators about performance impacts of patches to the operating system, hardware, etc.
- If information on code patches (especially optimizations) is also integrated to the unified database in an appropriate high-level description, it might be possible to state on their impact to performance based on data analytics. This will be valuable to decide on further similar modifications with respect to this and other codes.
- At runtime, it might be possible to analyze the first iterations of a job and deduce assumptions about the future behavior. Such information might be fed into the scheduler in order to

  - adapt priorities of jobs competing for a shared resource (favoring one might decrease his runtime but *not* increase the runtime of others)
  - run competing jobs with offset in order to allow for interleaved utilization of a shared resource

- Rating computational progress is a hard problem. This is due to the fact that performance is limited by inherent characteristics of an algorithm or application domain (such as ratio of required data and operations done on this data) while other algorithms and domains can achieve a better performance. Furthermore, algorithmic improvements might decrease metrics like Flop/s, memory B/W, IPC, etc. but decrease time to solution at the same time. It's unclear to us whether data analytics and/or artificial intelligence can help to tackle this problem. However, we plan to investigate this in the future. A starting point might be to automatically cluster similar jobs/codes in order to figure out those with a performance much worse compared to the mean of the cluster.

As mentioned above, the HLRS user support team proactively invites users to join optimization workshops and get continuous support. This is done according to their remaining computing time budget. Based on the unified database and analyses mentioned above, it will furthermore be possible to focus on applications with suboptimal performance, hence improving impact of efficiency enhancement activities.

Provide Performance Feedback to Users: For project admin purposes, users are able to check the status of their project via a web interface. This mechanism will be extended to give the project a "management level overview" (e.g. excellent performance, good performance, performance should be improved, lousy performance) based on the unified database. Using the same interface, more data will be provided and users with knowledge in programming and sufficient optimization skills may get information which section of the application should be improved (actual computation, communication, I/O, etc.). Of course, all users may request support to do so.

On the Detection and Interpretation of Performance Variations of HPC Applications

Dennis Hoppe, Li Zhong, Stefan Andersson, and Diana Moise

Abstract Supercomputers are synonymous with maximum performance, and thus one would expect that each run of an parallel applications would yield the same runtime provided that input parameters and data are unchanged. Practice, however, clearly demonstrates that this is not the case. Supercomputers are built with multi-user usage in mind, meaning that typically several hundred applications run simultaneously on a multitude of compute nodes. Although these compute nodes are assigned exclusively to users, network and data storage is shared among all; interferences between applications are inevitable. In this paper, we evaluate application runs on a Cray XC40 system. The objective is to identify so-called aggressor applications having a negative impact on the performance of simultaneously running applications resulting in unforeseeable longer runtimes. We discuss in this paper characteristics of aggressors and victims, as well as introduce several detection strategies to identify these victims, and thus also potential aggressors. Finally, a study demonstrates the effectiveness of the approach by identifying an aggressor and optimizing the source code, which resulted in less interference.

1 Introduction

High Performance Computing (HPC) is a key driving factor for both academic and industrial innovation. Its technology is well-established and actively applied in various areas of applications that are due to their complexity infeasible to be

D. Hoppe (✉) · L. Zhong
High Performance Computing Center Stuttgart, Stuttgart, Germany
e-mail: dennis.hoppe@hlrs.de; li.zhong@hlrs.de

S. Andersson
Amazon Web Services (AWS)
e-mail: lrande@amazon.com

D. Moise
Cray Inc., Basel, Switzerland
e-mail: dmoise@cray.com

© Springer Nature Switzerland AG 2020
M. M. Resch et al. (eds.), *Sustained Simulation Performance 2018 and 2019*,
https://doi.org/10.1007/978-3-030-39181-2_5

executed in a reasonable time on laptops or workstations. Common examples are climate modelling and weather predictions, crash tests, and pandemic simulations. HPC systems, however, are optimized for embarrassingly parallel tasks, and thus are first choice when executing these complex simulations. An HPC system serves simultaneously a multitude of users that perform a diverse set of applications on different groups of compute nodes. Although groups of compute nodes are assigned exclusively to users, applications still interfere with each other due to network traffic and access to shared file systems. These interferences can be severe, resulting in apparent performance variations; an application that runs usually for 10 h on 1000 compute nodes might require significantly longer to complete when affected. Since HPC centers usually have fees on a cost per node-hour, users attempt to keep costs low by minimizing required node-hours. As a result, users reserve compute nodes only for the required runtime of an application. If an application has not yet finished in the given time, progress made and data not saved is lost. Thus, HPC centers should guarantee that interferences with other applications running simultaneously on the large-scale system are kept at a minimum or, even better, be avoided altogether.

We face several challenges while trying to identify potential harmful applications that induce performance variations onto other applications. Ideally, an algorithm returns a set of affected applications, where each one is associated with a set of potential harmful applications or vice versa. Due to the fact that harmful applications are not impacted by performance variations, they do not emit any irregular characteristics to be identified. Harmful applications, which we will from now on call aggressors, are therefore hidden within the deep lake of all applications ever executed on the system, and thus are not readily detectable. Taken another approach, we are able to identify—by means of pure statistical methods such as outlier detection—thousands of potential affected applications, so-called victims. However, finding an outlier only indicates that a given application took sometimes significant longer to complete as on average. Another issue is that finding victims is therefore anything but an indicator for a reason why the application took longer than expected. Thus, we are in a poor starting situation: We would like to find aggressors and victims likewise, but we have a priori no insights about features to be looking for. And even if we would assume a given set of features (such as outliers), we would be limited to identify victims based on the assumed characteristics, likely missing victims that are affected in a yet unknown way. Another issue with using pure statistical methods is that initial experiments return a set of thousands of potential victims, which is infeasible to be manually analyzed by domain experts.

That remainder of this paper is organized as follows: Sect. 2 gives a focused overview about state-of-the-art in the domain of performance analysis and HPC job variations, while Sect. 3 then presents an exploratory analysis of past and current jobs ran on HLRS's Hazel Hen. Section 4 describes the methodology to identify job variations on production systems; identification strategies are discussed based on pure statistical methods and more contemporary AI techniques such as machine learning based. Section 5 then continues with the experiment setup and first results obtained in a small but representative study. Finally, Sect. 6 concludes with a summary and an outlook.

2 Related Work

The state of the art in evaluating performance variabilities in HPC workloads can be classified into three main approaches:

1. conducting experiments on a production system or test bed using real applications [1, 2],
2. evaluating the behavior of HPC workloads by means of benchmarks running on a production system or test bed [3], and
3. simulation of the underlying interconnect to predict characteristics of competing applications [4].

It should be noted that experiments in the latter two cases are, to the best of our knowledge, always performed in an isolated manner to avoid any further interference with real applications running simultaneously. In comparison to related work presented next, this paper focuses solely on the identification of real-world application performance variabilities in a production environment.

2.1 Evaluation of Performance Variations Based on Benchmarks

Skinner and Kramer [3] are one of the first to raise the issue of performance variability in HPC. The authors claim that without a proper understanding of the performance variability of a given application on a specific infrastructure, effective performance cannot be measured. Thus, Skinner et al. investigate several causes for performance degradation such as resource contention, MPI message sizes, OS jitter such as kernel process scheduling, overall system activity (e.g. monitoring), and finally so-called cross-application contention. Experiments were performed on four different HPC systems including a Cray T3E and an IBM SP with our 2500 sample runs using two benchmarks: LU and FT from the NAS Parallel Benchmarks (NPB) suite. Here, the coefficient of variance ranges from 2.62% to 15.58% for LU, and from 1.07% to 11.33% for runs of FT; the most performance variability was revealed on the Cray T3E system. Skinner et al. conclude that resource contention (e.g. over shared network paths or access to shared storages) and process scheduling are the main causes for performance variability. Furthermore, the authors suggest the use of monitoring tools to enable fine-grained profiling of applications. Still, they are aware of the fact that this is an impractical solution for production systems, where maximum performance is required at all times.

2.2 Evaluation of Performance Variations Based on Real-World Applications

Bhatele et al. [1] investigate potential causes for performance degradations of parallel application on HPC systems; causes include, besides the interference of parallel applications due to using the same network, OS jitter, job placement, and possible contention while multiple applications access a shared parallel file system such as Lustre. Bhatele et al. argue that performance variability of HPC jobs does not only impact the development of HPC applications (e.g. by making it more difficult to reproduce code changes with respect to performance), but also the overall resource allocation (e.g. estimating required resource reservation times). Experiments performed by Bhatele et al. on a Cray XE6 system revealed that the execution times of applications can range from 28% faster to 41% slower than the average observed performance. If conflicting jobs use the interconnect heavily, meaning that they are communication intensive, then performance degradations are likely to occur on a Cray XC40 system. This is in stark contrast to results obtained on IBM Blue Gene systems, which always assign a private network exclusively to a job, which is a trade-off between preventing job interferences and lower overall resource allocation. The work also concludes that job interferences are the dominant factor to performance variability.

Groves et al. [2] evaluate the impact of network traffic on the performance degradation of competing workloads. Specifically, the work reviews a subset of 25 Aries counters obtained per node and router. Since the authors rely on using PAPI to collect those counters, counters are limited to allocated compute resources for the currently running job. As a result, this approach does not obtain a global view of the entire infrastructure, and thus lacks accuracy. Furthermore, Groves et al. focus on a single application—MPI Allreduce. Allreduce is an MPI operation often used by HPC applications. It is shown that Allreduce is sensitive to noise, and therefore it is deemed to be appropriate to study the effects of performance variability of HPC jobs. Groves et al. evaluate the performance of Allreduce while running on up to 512 compute nodes; results show that Aries counters such as *flit* and *stall* are a good index of performance degradation of competing workloads. Still, these counters accounts for 70% of the variability in the slowest 10% of Allreduce communication. Although results of this paper are promising, the focus on a single application is not representative for the wide spectrum of complex applications running on HPC systems. Furthermore, the selection of PAPI to collect Aries network counters results in being limited to fabric counters previously assigned to a given job. As a consequence, it is not possible to obtain a global view of the entire system while trying to evaluate performance degradation. The work by Groves et al. was performed on a production system (Cray XC40).

2.3 Relying on Simulations to Identify HPC Workload Variabilities

Bhatele et al. [5] simulate network throughput for competing parallel workloads using the simulation tool Damselfly [6]. Here, simulation allows to circumvent a key challenge immanent with monitoring and evaluating real systems: calculating link traffic of individual jobs in a shared environment; well-known network counters such as Cray Aries counters do not differentiate between traffic of different applications at the link level. As a consequence, gathering network throughput on a real system always results in aggregated measurements per counters. Furthermore, simulating network traffic renders the need for having installed a monitoring tool at each link, router, or node as obsolete. Having (fine-grained) monitoring support for a HPC production system is not common: maximum performance is key, and thus each extra utility running in the background degrades performance per se. Bhatele et al. study interferences of parallel workloads by means of five different communication patterns implemented with MPI. Experiments on up to 131,072 cores were performed using a mix of five different MPI communication patterns to simulate parallel workload. Average and maximum traffic per job is then compared for different types of links (green, black, and blue); results obtained are then set side by side with runs of individual jobs. Whereas Groves et al. focus on the identification of network counters that could be good indicators for network congestion and thus performance degradation, Bhatele et al. analyze the impact of changes of the network topology as well modifications to job scheduling, job placement, and routing policies. The authors show that—depending on the communication pattern—either inter-group (i.e. blue) or intra-group (i.e. green) links are likely to cause congestion; black links do not have any impact on performance degradation. Although Bhatele et al. demonstrate that simulation can help to make design decisions during procurement, installation and configuration of supercomputers, we deem that simulation of a small set of communication patterns is not adequate to represent real-world HPC applications, which are often composed of mix of communication patterns. Nevertheless, results obtained propose to have a more detailed look at green and blue links when dealing with Dragonfly typologies.

2.4 Summary

When reviewing literature on performance variations, it becomes obvious that the Cray, and in particular also the Aries interconnect with its Dragonfly topology, is linked more often to interferences between simultaneous running jobs than other interconnects [1–3]. The main reason mentioned in several publications for performance degradation is the network interconnect and the resulting network congestion [1–3, 5]. At the High Performance Computing Centre (HLRS), we also observe these performance variations on a Cray XC40 system, which are described and analyzed in more detail in the remainder of this paper.

3 Insights into Hazel Hen

The goal of this work is to develop a generalized method which can detect and identify applications that have negative impacts on other applications on HPC. However, due to the large amount of applications running and the high cost, a try-and-error method is not practically impossible on the production system. Thus, a question arose: what type of HPC jobs impact the performance of other jobs?

3.1 Overview of Hazel Hens Jobs

The dataset which is utilized to support this study is consisted of Hazel Hen logs that are collected on the Cray System Management Workstation (SMW) in 2016. The Hazel Hen super computer, which is a Cray XC40 system and composed of more than 7700 compute nodes interconnected through Aries, is hosted at HLRS. Each diskless compute node of Hazel Hen has two sockets with an Intel Haswell processor, yielding nearly 185,000 cores. Furthermore, each compute node has 128 GB of random access memory (RAM) installed [7].

The log files used for study contain necessary information regarding all the applications running on the machine. However, the information contained in the Hazel Hen logs is quite sparse, namely: user id, job id, commandline, start time, end time, runtime, node list and the energy consumption. Other logs containing network traffic information and consumer information are currently not accessible. The total number of HPC application logs is 3,695,241, starting from "2016-01-01 18:58:18 CET" to the end time of log records at "2016-12-31 20:57:37 CET"; it originates from 220,225 unique jobs executed by 539 users.

Table 1 reveals the summary information about the energy consumption, number of nodes and runtime of all Hazel Hen jobs. In Fig. 1, the users are grouped into several categories by the number of submissions they have made in 2016. The users who submitted above 100,000 submissions are assumed to be administrators, and most of the actions they made are system operations. However, the system operations cannot be ignored in this study though the number is extremely large, as these actions are also probably the reason to cause performance variances.

Figure 2 shows the average number of jobs run in each day of the week. As one would expect, the number of events are higher throughout the work days and slightly less during the weekends. For the weekdays, although the jobs run on Tuesday

Table 1 Statistics of Hazel Hen's logs in 2016

Id	Description	Mean	Max	Min
0	Number of nodes	1.04	6656	1
1	Energy consumption	15,884,460	105,376,900,000	37
2	Runtime (min)	1115.6	630,136	0

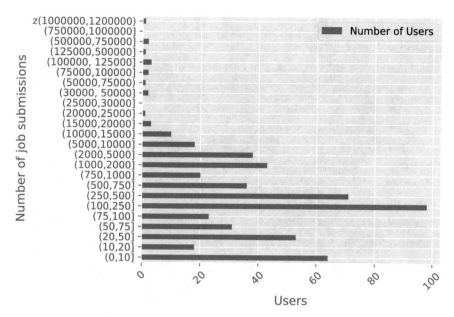

Fig. 1 User groups by job submissions

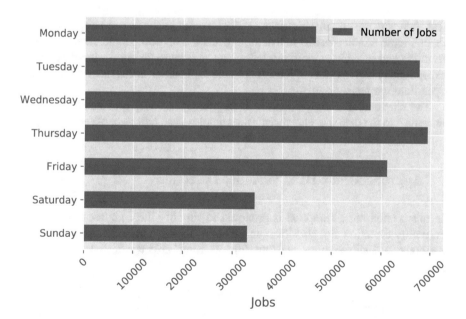

Fig. 2 Average number of jobs by days of the week

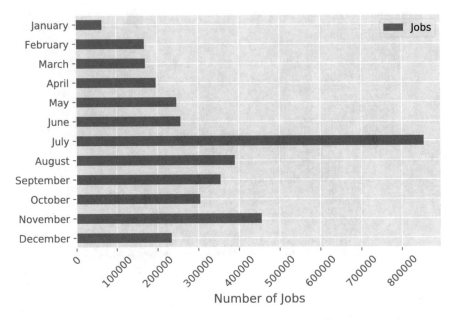

Fig. 3 Number of jobs by month of a year

and Thursday are slightly higher than other workdays, there is no big gap among them. Figure 3 indicates that unlike the expectation of most people, July, beginning from the end of which month most Germans choose to go out for holiday, has an exceptionally high number of jobs. A possible explanation is that most of the system operations are made on July.

The phenomenon observed above is helpful: it not only provides the intuitive insights into the data, but also reveals the complex correlations between jobs.

4 Approaches for HPC Variance Detection

In this section, an assumption is proposed to simplify the complex problem into an outlier detection and classification problem. To solve these two problems, two methods: "statistic analysis" and "machine learning" are investigated.

4.1 Assumptions

The initial assumption is that some potential harmful applications induce the performance variations on other applications. Ideally, an algorithm could be developed to return a set of harmful applications, where each one is associated with a set of potential harmful applications or vice versa. To better understand this problem, the following terminology is developed:

Victims *HPC jobs/applications that show high runtime variability.*

Aggressors *HPC jobs/applications potentially causing the variability.*

Understanding the nature of victims and aggressors is crucial for developing a detection strategy. Then the problem cloud be stated as:

Given an HPC job $x \in X$, for each job x, retrieve a set of all identical jobs that were already executed in the past with the same configuration and allocation of resources, X_{hist}.

Assume that x is a victim, x_v. Find the set of HPC jobs $Y = x \in X$ composed of potential aggressors. If $Y = \varnothing$, then declare x as an aggressor, x_a. Then identify features that label x either as a victim or as an aggressor. And classifying x as a victim is straightforward: $Y \neq \varnothing$.

A straight forward strategy is to identify the aggressors, However, when trying to follow this strategy and directly detect the aggressors, there exist several challenges. Due to the fact that the harmful applications' performances are not impacted by others, they do not emit any irregular characteristics to be identified. Aggressors are hidden within the deep lake of all applications ever executed on the system, and thus are not readily detectable.

Therefore, the other approach is taken. As the victims always show high variances in their runtime, and statistic method and machine learning algorithms are powerful in outlier detection, thus we will explore the two methods to identify the victims.

4.2 Statistic Method

The target of this section is to describe the statistic method adopted in this study to detect the victims. As a victim has the characteristic of showing great variances in its runtime. Therefore, those applications running on Hazel Hen, whose runtime has great differences from others, are regarded as outliers. Then, the problem of detecting victims is converted to outlier detection.

An Outlier is statistically defined as: One that appears to deviate markedly from other members of the sample in which it occurs [8]. A common approach to detect outliers from a data set consists in using the interquartile range (IQR), which is defined as the difference between the third and first quartile [9]. A quartile, is value which divides the dataset into four equal groups according to the distribution of values of a particular variable. The first quartile Q1 is defined as the middle number between the smallest number and the median of the data set, and the second quartile (Q2) is the median of the data. The third quartile Q3 marks the beginning of the last 25%. Generally, every value being smaller than Q1 − 1.5 IQR or larger than Q3 + 1.5 IQR is regarded as outlier [10].

Based on the simplification of this problem, the approach can then be divided into three main steps:

1. Data Aggregation: Different logs from different sources need to be collected and combined together. After that, the initial data we have at hand needs to be preprocessed so as to keep only the useful, correct information, the operations includes data combination (e.g.: identify logs of the same job and combine them), data clean (e.g.: remove null value, remove bad lines, discard lines with incomplete commands, disregard application runs that are too short probably because of errors, etc.), and so on.
2. Explorative Analysis: The goal of this step is to summarize the main characteristics of the data with visual and statistical methods. With this manipulation, it is intuitive to see what the data can tell us beyond the formal modeling or hypothesis testing task. Figures 1, 2, and 3 are generated from this step.
3. Outlier Detection: In this step, the changing geographies of point patterns based on runtime is examined to detect the outliers. The idea is to place a symmetrical surface over each point and then evaluating the distance from the point to a reference location based on a mathematical function and then summing the value for all the surfaces for that reference location [11]. During this process, IQR stated above is adopted to detect the outliers.

Through the procedure described above, the victims hidden behind the large amount of HPC tasks are identified and can thus be passed for further processing. However, although this method is quite effective in detection of victims, it can only be done on a single HPC job. For the huge variety of HPC jobs, it is too time and cost consuming to do this manipulation on the jobs one by one. Thus, a faster and more practical technique which enables the batch operation shall be developed.

4.3 Machine Learning

An alternative approach is to rely on machine learning. Machine learning, specifically classification, enables to train a model based on a given set of features and given target classes on a given set of data, and then apply this model to predict target classes for previously unseen data. Employing classification then would allow to consider a large set of features compared to the pure statistical outlier detection approach. Although this approach might be more suitable for the given problem statement, it comes again with some drawbacks. Classification algorithms such as SVM or Random Forests associate classes to data using complex statistics, and thus it is no longer reproducible what decisions were made by the algorithm to yield the given results. Moreover, such a classification might find victims and aggressors with high accuracy, but we still do not know which features or feature combinations are responsible for detecting an application as a victim or an aggressor. Feature ranking methods could be applied, but those would just rank the most important features that a model uses. Even if the algorithm declares that an application is an aggressor, we can neither be certain that the application is actually harmful nor do we know how this application interferes with other applications. Finally, training classification models require both a training set with manually labeled classes to be learnt. This is not given, and thus directly applying classification algorithms is not applicable.

When dealing with the unlabeled data, unsupervised algorithms such as an autoencoder has proven its great performance. An autoencoder [12] is an artificial neural network, which performs basically a dimensionality reduction on the given data. Typically use cases for autoencoders are image search, data compression, and topic modeling. This approach can be used, for example, to identify victims based on the fact that an application mostly behaves similar, but sometimes has different characteristics (which are yet unknown to us).

There are a variety of clustering models which are always jointly applied together with autoencoder, e.g. K-means [13], DBSCAN [14], OPTICS [15] and so on. In this study, K-means is experimented. K-means is a method of vector quantization, which aims to partition n observations into k clusters in which each observation belongs to the cluster with the nearest mean, serving as a prototype of the cluster. [16]

After the clustering of the Hazel Hen jobs, visualization of the 2 clusters can help check the quality of the clustering. For this purpose, T-distributed Stochastic Neighbor Embedding (t-SNE) is implemented. t-SNE can visualizes high-dimensional data by giving each data point a location in a two or three-dimensional map. And t-SNE is better than existing techniques at creating a single map that reveals structure at many different scales [17].

This approach can be summarized into three main steps:

1. Apply autoencoder to do dimension reduction and feature extraction
2. Use K-means to do clustering
3. Visualize the cluster result in 2D and 3D space
4. Go back to step 1 and adjust the model

Figure 4 shows the result of clustering for Hazel Hen jobs in a 2D space and the visualization of clustering result in 3D space is shown in Fig. 5. The blue point represents the tasks which are not affected, while the red points represent the victims.

4.4 Aggressors Detection

Sections 4.2 and 4.3 described two strategies for detecting victims. Standing on the victims we get by these two manners, all the applications which are running at the same time can be regarded as the potential aggressor set. To justify if a job is aggressor or not, the frequency of this job is taken as an important index, the higher the frequency, the higher probability it is an aggressor. Besides, two other significant features are paid attention to when analyzing, the number of nodes that a task runs on and the time overlap of the potential aggressors with the victims.

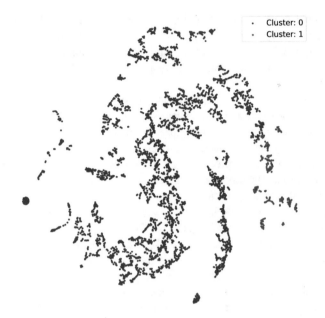

Fig. 4 Hazel Hen jobs into two clusters in 2D space

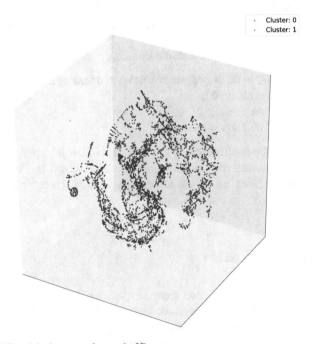

Fig. 5 Hazel Hen jobs into two clusters in 3D space

5   Experiment and Result

The experiment setup and implementation are described in this section. Further-more, the results we gained from experiments will be demonstrated and analyzed.

5.1   Experiment Setup

The experiment is conducted at the high-performance data-analytics (HPDA) sys-tem, a Cray Urika-GX [18] at HLRS. The HPDA system including x86 processors, Ethernet interconnects, and local disk-based storage solutions provide the basis for common data analytics clusters. Hadoop Distributed File System (HDFS), the dominant file system used in the domain of data analytics, is also enabled by local storage. On top of HDFS, various frameworks and tools are installed at HPDA system including Apache Hadoop [19] and Spark [20]. Apache Hadoop originates form the MapReduce paradigm, which always store the intermediate data on the disk, thus the read and write speed is accelerated. Besides, Spark enables in-memory processing. Consequently, the processing of data sets can be done in a much more efficient manner.

The statistics analysis and machine learning approaches are implemented with python, with utilization of libraries like sklearn, tensorflow, matplotlib, pandas, pyspark, keras and so on.

5.2   Implementation and Result

The analysis was run by CRAY on an initial dataset reflecting HLRS applications spanning 2 weeks. Based on it, 472 victims and 2892 potential aggressors were detected through this process. Among them, 7 of those potential aggressors were running on more than 1000 nodes and 3 of them were found repeatedly.

To further explore it, the test was also run on a larger dataset (SMW data over 3 months), and 3215 victims, 67,908 aggressors were identified by the same process, 17 of the 67,908 aggressors using more than 1000 nodes. The analysis took 268 s on 300 cores on the HPDA system.

Putting into a more practical environment, together with CRAY, a small test application called "canary" was created to run on Hazel Hen. The idea behind it is to create a practical application which enables the verification of a aggressor. It is continuously running an all-to-all test on 4 nodes of a cabinet group, which utilizes 64 bytes small messages and running only on the first socket of each node. Therefore, it is particularly sensitive to network interferences. When running on Hazel Hen, only one cabinet group is dedicated to running small jobs, and on the other 20 cabinet groups of Hazel Hen, a single canary job which uses 4 nodes is running.

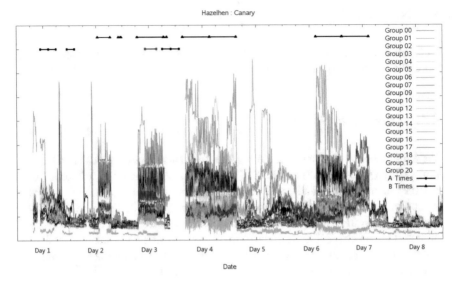

Fig. 6 Average number of jobs by month of a year

Figure 6 illustrates the performance of the canary, where x-axis represents the output timestamp and the y-axis represents the latency at a given timestamp of the 20 running canaries. As can be seen, the normal latency is about 100 μs, while a significant higher latency indicates the disturbance from aggressors. For example, two applications A and B were tested together with the canary output. The start- and end time of 6 A jobs are plotted on the $y = 1600$ axis (black and red line), where the red line indicates a job running very slowly. And the application B was running on a high number of nodes and plotted the start- and end-times of these jobs at $y = 1700$. It is quite clear that application B has a great influence on the latency of other canary tasks. Thus, application B is verified as the aggressor.

6 Conclusions and Future Work

Estimating the runtime of an HPC job is crucial for HPC customers and resource managers. The former has to plan ahead and reserve sufficient compute resources and time, whereas the latter requires an accurate prediction of resources becoming free to optimize the overall resource usage of a system. It is a challenging task, in particular for new HPC customers to assess the overall runtime of a job—it is more often than not trial and error. And yet, assessing the total runtime is critical with respect to cost spend on resources and potential data loss if a job should run longer than initially planned. Here, we observe, specifically at HPC centers, where hundreds of jobs run simultaneously and share the underlying network links, that unexpected runtime variations occur. In order to investigate the issue of HPC application performance variabilities, we rely on the methodology introduced in Hoppe et al. [7] to classify jobs as so-called victims and aggressors. The ultimate goal is then to identify aggressors in order to assist application owners to optimize their code to be less intrusive in future.

To achieve this ambitious goal, we introduced in this paper two strategies: a pure statistic approach and machine learning techniques to identify in a first instance potential victims. However, both have their limitations. Where the statistic method based on outlier detection successfully detects unusually long runtimes, it does not provide any further insights about why the job ran longer, and it also does not link the outlier to potential aggressors. The machine learning approach has the advantage to be based on features to be trained for a model to classify jobs into victims or aggressors. With the help of feature ranking methods, one would be able to identify the most relevant features that are characteristic for a victim or an aggressor. However, since no gold standard is available on a production system, applying supervised learning algorithms such as SVM or RandomForests is not applicable, and thus we have to rely on unsupervised methods such as Clustering techniques. With an ensemble method consisted of autoencoder, k-means and t-SNE, the tested Hazel Hen tasks are clustered into two groups, and visualized by red and blue points in 2D and 3D spaces, where the 428 red points represent the victims while the 4953 blue points stand for the 4953 unaffected jobs. Evidently, this method show its advantages over statistic method as it provides the ability to batch process data in a faster and more practical way. Still, due to the complicated inner network structure and data compression, it is far more difficult to justify and explain the decisions made by the algorithm.

Finally, to get better insights into characteristics of HPC jobs, a test application named "canary" was executed on a small cluster with the goal of better identifying aggressors. Based on the canary output, we started analyzing the workload and latency while running two test applications. With this process, the potential aggressors are successfully justified.

As future work, we foresee to include more log information into the analysis (e.g. Aries counters need to be link to individual jobs), as well as to investigate in more detail the potential of unsupervised methods to identify victims and aggressors. Performing this analysis on a production system limits monitoring capabilities, and on other hand, provides new opportunities to set up eventually an AI-based live job monitoring system for the identification of victims and aggressors.

References

1. Bhatele, A., et al.: There goes the neighborhood: performance degradation due to nearby jobs. In: Proceedings of the International Conference on High Performance Computing, Networking, Storage and Analysis (SC'13). Colorado (2013)
2. Groves, T., Gu, Y., Wright, N.J.: Understanding performance variability on the aries dragonfly network. In: International Conference on Cluster Computing (CLUSTER'17). Hawaii (2017)

3. Skinner, D., Kramer, W.: Understanding the causes of performance variability in HPC workloads. In: Proceedings of the International Symposium on Workload Characterization (IISWC'10). Washington (2005)
4. Bhatele, A., Jain, N., Livnat, Y., Pascucci, V., Bremer, P.T.: Evaluating system parameters on a dragonfly using simulation and visualization (2015)
5. Bhatele, A., et al.: Identifying the culprits behind network congestion. In: International Parallel and Distributed Processing Symposium (IPDPS'15). Hyderabad (2015)
6. Jain, N., et al.: Maximizing throughput on a dragonfly network. In: Proceedings of the International Conference for High Performance Computing, Networking, Storage and Analysis (SC'14). New Orleans (2014)
7. Hoppe, D., Gienger, M., Bönisch, T., Shcherbakov, O., Moise, D.: Towards seamless integration of data analytics into existing HPC infrastructures. In: Proceedings of the Cray User Group (CUG). Redmond (2017)
8. Grubbs, F.E.: Procedures for detecting outlying observations in samples. Technometrics **11**(1), 1–21 (1969)
9. Mood, A.M., Graybill, F.A., Boes, D.C.: Introduction to the Theory of Statistics. McGraw-Hill, New York (1974)
10. Benkert, K., Gabriel, E., Resch, M.M.: Outlier detection in performance data of parallel applications. In: 2008 IEEE International Symposium on Parallel and Distributed Processing, pp. 1–8. IEEE, Piscataway (2008)
11. Anderson, T.K.: Kernel density estimation and K-means clustering to profile road accident hotspots. Accid. Anal. Prev. **41**(3), 359–364 (2009)
12. Ng, A.: Sparse autoencoder. CS294A Lect. Notes **72**(2011), 1–19 (2011)
13. Krishna, K., Murty, N.M.: Genetic K-means algorithm. IEEE Trans. Syst. Man Cybern. Part B Cybern. **29**(3), 433–439 (1999)
14. Schubert, E., Sander, J., Ester, M., Kriegel, H.P., Xu, X.: DBSCAN revisited, revisited: why and how you should (still) use DBSCAN. ACM Trans. Database Syst. **42**(3), 19 (2017)
15. Ankerst, M., Breunig, M.M., Kriegel, H.P., Sander, J.: OPTICS: ordering points to identify the clustering structure. In: ACM Sigmod Record, vol. 28(2), pp. 49–60. ACM, New York (1999)
16. Jain, A.K.: Data clustering: 50 years beyond K-means. Pattern Recogn. Lett. **31**(8), 651–666 (2010)
17. Maaten, L.V.D., Hinton, G.: Visualizing data using t-SNE. J. Mach. Learn. Res. **9**, 2579–2605 (2008)
18. Cray Inc.: Cray Urika-GX product brochure. http://www.cray.com/sites/default/files/Cray-Urika-GX-Product-Brochure.pdf
19. White, T.: Hadoop: The Definitive Guide/Tom White. O'Reilly, Farnham (2012)
20. The Apache Software Foundation: Apache spark—lightning-fast cluster computing. http://spark.apache.org/

Using the NEC Aurora TSUBASA for High-Order Discontinuous-Galerkin in Ateles

Harald Klimach and Sabine Roller

Abstract High-Order schemes are attractive for modern computing systems as they can achieve accurate representations of solutions with relatively little memory. In this work we look into the usage of the *NEC Aurora TSUBASA* system for a high-order discontinuous Galerkin solver. The *NEC Aurora TSUBASA* system is a vector architecture that is transparently combined with a scalar host system. We explore this unique environment for development and execution, as well as providing some first performance observations with our discontinuous Galerkin solver *Ateles*.

1 Introduction: High-Order Discontinuous Galerkin

One of the limiting factors in modern computing is the memory and memory bandwidth of a system. Due to the different rate in growth of computing operations and memory operations, access to memory is by now slow in relation to the computations [1]. Numerical algorithms with small memory footprint and bandwidth requirements are, therefore, attractive for efficient computations on modern systems. High-order schemes offer high quality representations of solutions with few degrees of freedom. However, there is a strong interaction between all the degrees of freedom in the high-order approximation, which hurts parallel computations. This can be addressed by decomposing the computational domain into individual elements that are coupled less tightly. The discontinuous Galerkin scheme is a method that enables exactly this setup. A mesh with non-overlapping elements is used to discretize the computational domain and the solution in each element is approximated by some function, usually a polynomial expansion series. Between elements only surface fluxes need to be exchanged. In our implementation of this method in the solver *Ateles*, we use Legendre polynomials as a basis to

H. Klimach (✉) · S. Roller
University of Siegen, Siegen, Germany
e-mail: harald.klimach@uni-siegen.de; sabine.roller@uni-siegen.de

© Springer Nature Switzerland AG 2020
M. M. Resch et al. (eds.), *Sustained Simulation Performance 2018 and 2019*,
https://doi.org/10.1007/978-3-030-39181-2_6

represent the solution and transform this modal representation into nodal data for nonlinear terms.

The data within single elements is strictly structured as the multidimensional polynomials are obtained by a tensor product of onedimensionals. This opens the possibility to use a dimension-by-dimension strategy in the numerical algorithms, greatly reducing the computational effort. In contrast the mesh of elements is organized with greater flexibility. *Ateles* utilizes an octree representation, but the overall geometrical shape of the considered computational domain may have an arbitrary shape and explicit neighbor information is obtained from the octree structure. This setup is not too different from block-structured meshes when considering the data organization. It enables the flexibility of unstructured meshes while also providing chunks of highly structured data. For a vector system like the *NEC Aurora TSUBASA* a viable strategy then emerges from vectorizing the operations within elements, while maintaining all the flexibility in the outer unstructured mesh.

The total number of degrees of freedom within an element (per state variable) in three dimensions is given by the polynomial degree plus one cubed. In general the number of degrees of freedom (N) for an approximation with a polynomial series of maximal degree q in d dimensions is given by:

$$N = (q + 1)^d \tag{1}$$

While there are few operations that can be done independently for all the degrees of freedom, most algorithms have a dependency in one dimension. This is a direct consequence of the dimension-by-dimension strategy followed in the tensor product representation of the polynomials. Thus, the number of independent degrees of freedom available for straight forward vectorization (M) is found by

$$M = (q + 1)^{d-1} \tag{2}$$

We are here aiming for a high-order with $q > 11$. This is a threshold that we found *Ateles* to work most efficiently at on typical scalar *x86* systems. With a polynomial degree of $q = 15$ there is a sufficient number of independent degrees of freedom to completely fill the vector data registers of the *NEC Aurora TSUBASA* that have a length of 256. This high-order discontinuous Galerkin scheme should, therefore, nicely fit the vector system and we report some first experiences with this newly available system.

While the mesh is organized unstructuredly with an arbitrary shape of the computational domain, the state variables are stored in a single array for all elements with the same internal structure. This array has three dimensions. The first dimension represents the elements. The second dimension represents the degrees of freedom describing the multidimensional polynomial and the third dimension represents the different state variables of the equation system. For the interaction between elements surface data is required. This is in a secondary data structure and a mapping from the volumetric to the surface data is required.

2   Ateles Software Environment

Ateles is an application mostly written in modern Fortran. The largest part is actually restricted to the Fortran 95 standard, however it heavily relies on the ISO-C-Binding introduced in Fortran 2003. Some other features from the Fortran 2003 standard are also utilized, but the code tries to avoid object oriented features as they were not deemed sufficiently stable in the compilers for a long time. Features from the Fortran 2008 standard are only used in optional functionality that can be switched off at compiler time, if the compiler does not support them.

Configuration and compilation is realized with the *waf* [2] build tool that requires Python. *waf* can detect the compiler of the platform automatically during the configuration phase for the build process. The necessary script for the automatic adaption was easily obtained by adapting existing detectors. Support for this automatic detection of the *nfort* compiler was added to the official *waf* in release *2.0.14*. It should be noted that the automatic configuration of the project may utilize compiling test executables and running them on the fly to detect available features. With the transparent execution model of the *NEC Aurora TSUBASA* this poses no problem at all.

To pre-process the Fortran code we make use of the conditional compilation program *CoCo* [3] from Daniel Nagle that implements the preprocessing once described in the Fortran standard as an extension. For the description of settings *Ateles* uses *Lua* scripts and, thus, relies on this library. It is written in C and built automatically within *Aotus*, our Fortran library that encapsulates the interaction with *Lua* via the ISO-C-Binding available with the Fortran 2003 standard.

Other than these tools there is only a dependency on an *MPI* library and optionally the *FFTW* may be utilized, but this is not considered here.

3   Porting to NEC Aurora TSUBASA

The *NEC Aurora TSUBASA* system provides a scalar *x86* vector host that takes care of system calls like IO. Attached to this vector host are one or multiple vector engines that provide vectorized computations with a vector length of 256. Executables compiled for the vector engine are loaded onto them from the vector host and executed there transparently. Each vector engine provides four cores and the system installed at *ZIMT* at the University of Siegen is equipped with two vector engines.

As described above in Sect. 2 there are only few dependencies required to build the solver. Compilation of the Fortran 2003 code base worked straight away without any issues with the *nfort* compiler in version 1.6. However, when running our testsuites we encountered some problems. We use *waf* to perform unit tests during compilation. Thanks to the transparent execution this works without a hassle on the *NEC Aurora TSUBASA* system with the restriction that executables that call

Listing 1 Loop broken by aggressive optimization

```
do orig=1,nModes
  sign_factor = mod(orig,2)*2 - 1
  do m=1, orig-1
    split_matrix(orig, m) = sign_factor * split_matrix(m,
        orig)
    sign_factor = -sign_factor
  end do
end do
```

Listing 2 Split loop avoiding problems with aggressive optimization

```
do orig=1,nModes
  sign_factor = mod(orig,2)*2 - 1
  !$NEC ivdep
  do m=1, orig-1, 2
    split_matrix(orig, m) = sign_factor * split_matrix(m,
        orig)
  end do
  !$NEC ivdep
  do m=2, orig-1, 2
    split_matrix(orig, m) = -sign_factor * split_matrix(m,
        orig)
  end do
end do
```

MPI_Init also need to be called with mpiexec, even if only a single process is used for the execution. Some of these unit tests failed. One minor issue that we found was that the spacing function for Fortrans tiny values of reals did not work properly. Another was that a loop with alternating signs was broken by more aggressive optimizations than level 1. Both issues got resolved by later compiler versions.

The loop construct that got broken by more aggressive optimizations is shown in Listing 1.

The alternation of the sign is better achieved by splitting the loop as shown in Listing 2. With the split loop the aggressive optimization does not cause any other issues in our unit tests.

The gamma function available in the Fortran 2008 standard is provided by *nfort*. It provides minimally different values than *GCC*'s *gfortran* within numerical accuracy.

Aside from the unit tests run by *waf* during compilation, we maintain a testsuite with setups to check the results of overall program execution. For those we did not find any variations from the reference results on the *NEC Aurora TSUBASA* even when using the most aggressive optimization flags. All in all the porting did not prove difficult and we found the system with its transparent execution on the Vector Engines to be convenient for the building and testing process.

4 Tuning for the Vector System

There are two main classes of equation systems that we need to consider for the performance of *Ateles*. The first class are linear equation systems like Maxwell or linearized Euler. They can be computed in a purely modal approach and their most important computational work load is the multiplication with the mass and stiffness matrices. With the chosen polynomial basis these matrices can be computed efficiently without increased costs per degree of freedom for higher order discretization. The second class are nonlinear equation systems like the Euler equations for compressible inviscid flows. For those systems a transformation to a nodal basis is required for the computation of the nonlinear terms [4]. This transformation gets increasingly expensive in terms of necessary operations with increasing polynomial degree. With the dimension by dimension approach we follow, we expect the required operations per degree of freedom to grow linearly with the polynomial degree of the discretization. As a comparison for the performance we will use the cluster system *Horus* installed at *ZIMT* at the University of Siegen. It is equipped with two 6 core *Intel Sandy Bridge (Xeon X5650)* processors in each node operated with 2.66 GHz. As a performance measure that allows us to compare the performance for different scheme orders and mesh sizes, we use a measure of million degree of freedom updates per second (*MDUPs*). The more degree of freedom updates per second we can achieve, the faster our simulations will be computed.

4.1 Linear Equation System

Figure 1 shows the single core performance for the linear Maxwell equations achieved on the *Aurora* system in comparison to the performance on the *Horus* system without tuning for the vector system.

It should be noted, that the code actually had already been somewhat vectorized for the *NEC SX ACE*. The compiler directives where easily converted for the new *nfort* compiler on *Aurora*. Other than that there was no further tuning effort for this run. We can observe a relatively high single core performance on the *NEC Aurora TSUBASA* with the code as it is, but there are strong degradations in the performance for specific scheme orders. These are most pronounced for multiples of 16 but also for smaller powers of two we can observe a degradation. To analyse the nature of this performance degradation, we look into the *ftrace* for the run with a scheme order of 48. The most important routines for this run turns out to be the projection

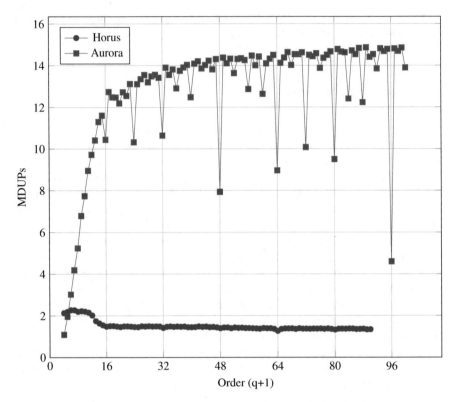

Fig. 1 Serial performance for the Maxwell equations with 64 elements on Aurora compared to Horus

Table 1 Excerpt from the tracing of *Ateles* for Maxwell equations and a discretization with polynomials of degree $q = 47$ and 64 elements

Procedure	MFLOPS	V. OP%	V. LEN	Port confl.	Stride
PrjFluxZ	636.7	99.97	230.3	0.571	$64 \times 48 \times 48$
PrjFluxY	3845.7	99.97	230.3	0.125	64×48
PrjFluxX	6110.8	99.97	230.3	0.048	64

of the fluxes to the test functions. This needs to be done for each direction, which come with different strides in the memory access.

Table 1 shows the *ftrace* information for this routine with the three different directions. It shows that the vector operation ratio and the average vector length is reasonable high, but we get long times spent in port conflicts for the Z and Y directions where strides are multiples of 1024. We could avoid this by introducing some paddings, but it can as easily be avoided by increasing the scheme order if a stride of 1024 would be encountered. Table 2 shows the performance for the same three routines when a scheme order of 49 is used instead of 48. Here the strides are not multiples of 1024 and the observed port conflict times are small for

Table 2 Excerpt from the tracing of *Ateles* for Maxwell equations and a discretization with polynomials of degree $q = 48$ and 64 elements

Procedure	MFLOPS	V. OP%	V. LEN	Port confl.	Stride
PrjFluxZ	7000.0	99.97	240.1	0.075	$64 \times 49 \times 49$
PrjFluxY	6893.9	99.97	240.2	0.075	64×49
PrjFluxX	6438.5	99.97	240.4	0.047	64

all three directions. In this case the computationally most expensive routine is the multiplication with the inverse of the mass matrix.

Aside from the performance degradations due to the bank conflicts with bad memory access strides, we find that the number of degree of freedom updates increases with the order and for scheme orders above 16 we find reasonable performance. A scheme order of 16 is the point where the vector length of 256 can be completely filled by most of the algorithms. On the scalar system of *horus* we see the opposite trend, where the performance decreases with increasing scheme order. A higher performance can be observed where the elements are sufficiently small to fit into the cache, up to a scheme order of 11–12. For higher scheme orders the performance drops significantly up to scheme order 16 and afterward keeps on slowly declining. It can also be noted, that the performance slightly decreases for multiples of 16 in the scheme order, similar to the behaviour on the *NEC Aurora TSUBASA*.

4.2 Nonlinear Equation System

We use the Euler equation system for inviscid compressible flows to assess the performance of the *Aurora* for nonlinear equations in *Ateles*. The performance without any tuning is shown in Fig. 2. For the nonlinear equation system we expect increasing operations per degree of freedom update and, therefore, decreasing degree of freedom updates per second with increasing scheme order. For the scalar system in *Horus* this can be nicely observed after the initial cache region with increased performance. Similarly a slight decrease in the performance can be observed on the *NEC Aurora TSUBASA* system, though it is less pronounced and only takes effect for scheme orders larger than 40. Peak performance is achieved for a scheme order of 39.

Similarly to the behaviour for the linear equations we observe a drop in performance for multiples of 16 in the scheme order. Though the effect is less pronounced. Again we look at the *ftrace* for the run with a scheme order of 48. The most important routines are shown in Table 3. Similar to the linear case we note that the projection of the flux appears here and we find a large time spent in port conflicts. But the inverse of the mass matrix is nevertheless taking most of the compute time with a low floating point operation rate. This is due to the short vector length of 8 that is utilized in this routine. Obviously the vectorization is here done

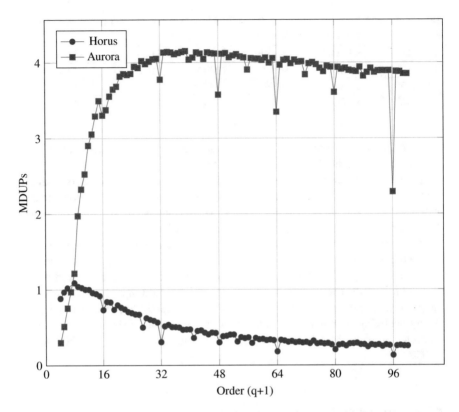

Fig. 2 Serial performance for the Euler equations with 8 elements on Aurora compared to Horus

Table 3 Excerpt from the tracing of *Ateles* for Euler equations and a discretization with polynomials of degree $q = 48$ and 8 elements without tuning

Procedure	MFLOPS	V. OP%	V. LEN	Port confl.
InvMassMat	257.4	60.88	8	0
Conv2Overs	0.0	46.18	17.6	13.19
PrjFluxZ	1403.0	99.97	230.2	14.89
L2 project	205,627.0	99.97	227.8	0

over the number of elements, not the degrees of freedom inside the element. The `Conv2Overs` is a pure memory copy routine that allows the use of more points in the nodal representation to avoid aliasing errors. No floating point operations are done in this routine. The final routine in this excerpt is the `L2 Project` that takes care of projection the modal space to the nodal space. It boils down to a matrix-matrix multiplication, which is recognized by the compiler and replaced by an optimized kernel, yielding the high sustained performance of 205 GFLOPs. This is the part that needs to be computed in addition for the nonlinear equation systems when compared to linear ones and is responsible for the increased computational effort for higher polynomial degrees.

Table 4 Excerpt from the tracing of *Ateles* for Euler equations and a discretization with polynomials of degree $q = 49$ and 8 elements without tuning

Procedure	MFLOPS	V. OP%	V. LEN	Port confl.
InvMassMat	257.3	60.91	8	0
L2 project	194,735.0	99.09	227.8	2.63
Conv2Overs	0.0	46.18	17.6	0.02

Listing 3 Array syntax with vectorization only over the first index

```
subroutine compute( nTotal, nDofs, nScalars, state,
    state_der)
  integer, intent(in) :: nTotal, nDofs, nScalars
  real(kind=rk), intent(inout) :: state(nTotal, nDofs,
    nScalars)
  real(kind=rk), intent(in) :: state_der(:,:,:)

  state = state + state_der(:nTotal, :nDofs, :nScalars)
end subroutine compute
```

Again, the port conflict issue can be avoided by using a different scheme order, where we do not end up with bad strides. In Table 4 the most important computational routines are shown for the run with a scheme order of 49. For this scheme order the port conflict times vanish and the projection of the fluxes consumes only a small fraction of the overall time.

The short vector length was a little surprising as we did not experience this on the *NEC SX ACE* system. However, it is easily resolved by collapsing the loop over elements and the loop of the independent degrees of freedom. With this collapse the average vector length for the 49th order simulation gets up to 252.7, allowing the code to actually utilize the vector instructions.

We also found short vectors surprisingly in array syntax statements with multiple dimensions with the compiler in version 1.6 as in the example shown in Listing 3. Even though all operations are completely independent we only observe a vector length given by the first index of the multidimensional array.

Using a manually collapsed loop instead with a compiler directive overcomes this shortcoming and we achieve an average vector length of 255.4 for this routine. However, please note that this shortcoming has been fixed in newer compiler versions.

After fixing the short loops we now achieve an overall performance of 67 GFLOPs for the 49th order scheme. With a theoretical peak of 268.8 GFLOPs for a single *NEC Aurora TSUBASA* core, this means a sustained performance of around 25%. Table 5 shows all the routines with more than 5% of the overall running time.

Figure 3 compares the performance with long vectors for scheme orders up to 100 compared to the variant without the tuning and short vectors. We now observe a performance that gets close to the linear equation system in its peak between spatial scheme order 32 and 50. With the improved performance also the degrading

Table 5 Performance of the routines with more than 5% of the overall running time for Euler equations with a scheme order of 49 and 8 elements

Procedure	Time (%)	MFLOPS	V. OP%	V. LEN	Port confl.
L2 Project	32.2	197,256.0	99.09	216.5	7.082
InvMassMat	10.8	4469.4	99.72	252.7	0.003
Vol2FaceY	7.8	1235.4	99.81	255.4	0.003
Vol2FaceZ	6.7	1448.8	99.80	255.4	0.003
Vol2FaceX	6.4	1502.1	99.80	255.4	0.001
Timest.RK4	6.0	827.8	99.77	255.4	0.001

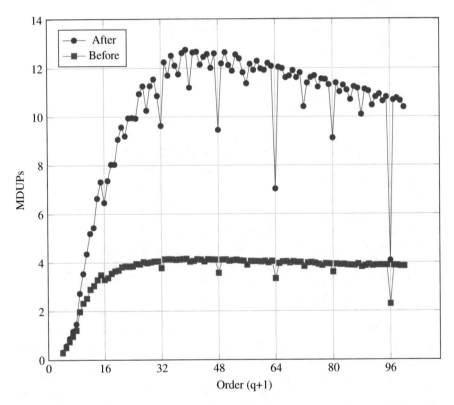

Fig. 3 Serial performance for the Euler equations with 8 elements on Aurora after fixing short vectors compared to before

number of degree of freedom updates for higher scheme orders, due to the additional operations gets more visible.

Figure 4 shows how the full *NEC Aurora TSUBASA* node with both vector engines and 8 cores compares to a single node of *Horus* with 12 *Intel Xeon X5650 (SandyBridge)* cores. For sufficiently high scheme orders, this single *NEC Aurora TSUBASA* node achieves a factor of 10 and higher in performance. Thus, the vector

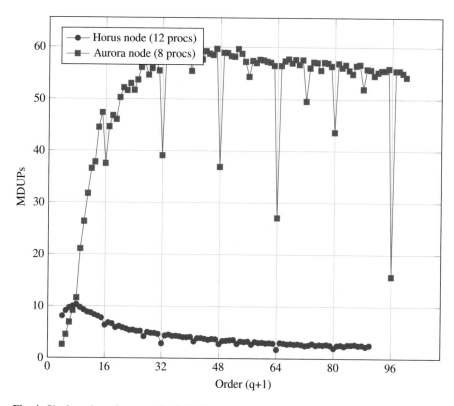

Fig. 4 Single node performance for the Euler equations with 64 elements on Aurora using 8 cores compared to a single node of Horus with 12 processes

engine is especially attractive for all settings with a high scheme order. In the case of high scheme orders we usually have only few elements, which limits the number of processes on which we can distribute the simulation. Strong but few processors with a high memory bandwidth appear to be an ideal match for the numerics of high-order discontinuous Galerkin schemes.

5   Conclusion

Ateles was easily ported to the *NEC Aurora TSUBASA* system. Only some minor quirks needed be overcome for the *nfort* compiler in version 1.6. The setup with the vector host and transparent execution on the vector engines proved to be very convenient for the deployment of the application. With high-order discretizations we achieve a high sustained performance on the vector system with an overall sustained performance of around 25%, yielding ten times the performance that we get from the scalar processors in the cluster.

Acknowledgements We thank the computing center of the University of Siegen (*ZIMT*) for providing the *NEC Aurora TSUBASA* system and Holger Berger of *NEC* for his kind assistance.

References

1. Hennessy, J., Patterson, D.: Computer Architecture: A Quantitative Approach, 5th edn. Morgan Kaufmann, Los Altos (2012)
2. Nagy, T.: Waf: the meta build system. https://waf.io. Accessed 12 July 2019
3. Nagle, D.: About the program CoCo. http://www.daniellnagle.com/coco.html. Accessed 12 July 2019
4. Anand, N., Klimach, H., Roller, S.: Dealing with non-linear terms in the modal high-order discontinuous Galerkin method. In: Resch, M., Bez, W., Focht, E., Patel, N., Kobayashi, H. (eds.) Sustained Simulation Performance 2016. Springer, Heidelberg (2016)

Performance Evaluation of SX-Aurora TSUBASA by Using Benchmark Programs

Kazuhiko Komatsu and Hiroaki Kobayashi

Abstract This paper evaluates the basic performance of the latest vector supercomputer, SX-Aurora TSUBASA, in order to clarify its potential. First, the memory bandwidth, which is one of the features of SX-Aurora TSUBASA, is evaluated by using the Stream benchmark. Next, the performances of the Himeno benchmark and the HPCG benchmark are examined. From these evaluations, it is clarified that the high sustained memory bandwidth can be achieved. It is also clarified that the high memory bandwidth of SX-Aurora TSUBASA can achieve the high sustained performance in other benchmarks compared to SX-ACE.

1 Introduction

Recently, supercomputers have been used not only in many fields of cutting-edge researches but also in various industrial fields such as engineers. Supercomputers have played an important role not only as research infrastructure but also as social infrastructure. Thus, the computational requirements for supercomputers are increasing. To meet such high computational requirements, the computing performance of supercomputers have been improved.

The multi and many-core technology, which integrates multiple or many cores into one processor, is used in various processors. In addition, vector technology that calculates multiple elements simultaneously by one instruction is widely adopted not only to vector processors but also to Intel-based scalar processors and accelerators such as GPUs. Due to these technological innovations, the computing performance of supercomputers has improved by about 2600 times over the past

K. Komatsu (✉)
Cyberscience Center, Tohoku University, Aoba-ku, Sendai, Japan
e-mail: komatsu@tohoku.ac.jp

H. Kobayashi
Graduate School of Information Sciences, Tohoku University, Aoba-ku, Sendai, Japan
e-mail: koba@tohoku.ac.jp

© Springer Nature Switzerland AG 2020
M. M. Resch et al. (eds.), *Sustained Simulation Performance 2018 and 2019*,
https://doi.org/10.1007/978-3-030-39181-2_7

15 years [1]. The theoretical performance of the world's highest performance supercomputer has reached 200 Pflop/s.

However, the improvement in memory performance is relatively low compared to that in computational performance of supercomputers, called as the memory wall problem. Thus, the gap between theoretical performance and sustained performance is increasingly widespread. In other words, only the computation-intensive applications can benefit from the high theoretical performance. The memory-intensive applications that require high memory performance cannot use the high theoretical performance of supercomputers, and then the sustained performance is limited by memory performance.

Vector supercomputers such as the SX series developed by NEC are known as supercomputers that have higher memory performance than the other super-computers. The vector supercomputers can achieve high sustained performance in applications that require high memory performance [2, 3]. Although vector super-computers are a minority in the list of TOP 500 that measures only computational capability, the vector supercomputers achieved high sustained performance in the HPCG benchmark, which is developed to be close to practical applications [4, 5]. In the HPCG benchmark, although the efficiencies of supercomputers equipped with scalar processors and accelerators are 1.0–4.8%, the efficiency of vector supercomputer SX-ACE reaches more than 10%.

As not only computational performance but also memory performance has attracted attention, a new vector supercomputer SX-Aurora TSUBASA has been released with the world's highest memory performance. SX-Aurora TSUBASA is a vector supercomputer that inherits and improves the advantages of the previous SX series.

SX-Aurora TSUBASA has been developed under two important concepts; high usability and high sustained performance. To meet these two concepts, a new system architecture and high bandwidth memory are adopted in SX-Aurora TSUBASA. For high usability, SX-Aurora TSUBASA adopts a new system architecture that consists of Vector Engine (VE) with vector processor and Vector Host (VH) with a standard x86 processor. Although it is similar to the traditional system architecture of accelerators, the concept is different. In SX-Aurora TSUBASA, whole applications are basically executed by VE while only OS functions such as system calls are offloaded to VH.

By these system architecture and execution model, there are two major advantages over conventional accelerators. The first advantage is that data transfer between VE and VH, which tends to be a bottleneck in the conventional model of accelerators, can be avoided. Since only data that is required for OS related functions is transfered between VE and VH, the data transfer of computation results can be reduced. As a result, the bottleneck by the data transfer can be avoided. The second advantage is that no special programming is required to develop an application for VE. As the conventional SX series, the compiler performs automatic vectorization and automatic parallelization. Thus, programs can be executed on vector processors without special programming.

In addition, SX-Aurora TSUBASA can provide a high memory bandwidth by integrating six HBM modules (high bandwidth memory) into a VE in cooperation with TSMC, Broadcom, and NEC. The six HBM module integration enables high memory bandwidth corresponding to the high computing performance of the multi-vector cores. As the high memory bandwidth and the high computational performance are balanced, it is possible to achieve a high sustained performance, especially in applications that require memory performance.

This paper evaluates the basic performance of the latest vector supercomputer SX-Aurora TSUBASA. First, memory performance, one of the features of SX-Aurora TSUBASA, is measured using the stream benchmark. Then, the basic performances of SX-Aurora TSUBASA are clarified compared with SX-ACE, Xeon Gold, Tesla V100, and KNL using the Himeno benchmark and the HPCG benchmark.

2 Overview of SX-Aurora TSUBASA

From SX-1 developed in 1983 to SX-ACE, the previous SX series of vector supercomputers have been pursuing high sustained performance, especially for applications that requires high memory bandwidth. Handing over the SX series of DNA, SX-Aurora TSUBASA, a new vector supercomputer equipped with a newly designed vector processor, was announced in October 2017. SX-Aurora TSUBASA is more efficient in terms of power and space than SX-ACE. SX-Aurora TSUBASA achieves 1/5 power consumption and 1/10 floor area at the same computing performance of SX-ACE.

The system architecture of SX-Aurora TSUBASA is different from the previous SX series. SX-Aurora TSUBASA consists of one VH and one or more VEs. By adopting standard Linux as the OS of VH instead of SUPER-UX, which is the original OS for the SX series, the usability of Linux is obtained. By migrating a usual Linux environment to VH, a programmer can use SX-Aurora TSUBASA in a familiar environment. In addition, an OS for VE called *VEOS* is executed on VH to control VE from VH.

VE is implemented as a PCI express card with a newly designed vector processor. Three types of VE are available depending on the frequency and memory capacity of the vector processor. SX-Aurora TSUBASA can be flexibly configured not only as large-scale supercomputers but also as personal computer. SX-Aurora TSUBASA has three product lines: A100 series, A300 series, and A500 series. A100 is a workstation model with a minimum configuration consisting of one VH and one VE. A300 is a standard rack model that uses air cooling. A300 can be integrated up to 8VE. A500 is designed for large-scale supercomputers that use the water and air cooling. A500 can be mounted up to 8VE per 1VH and up to 8VH per rack.

Fig. 1 A vector processor in a VE (provided by NEC)

2.1 Architecture of Vector Engine

A vector processor installed in VE is designed to achieve high sustained performance with appropriate power efficiency in applications that require memory bandwidth performance.

Figure 1 shows the CPU package of a VE. The VE processor in the shape of a rectangle is placed at the center of the package. Six HBM2 memories are arranged on both sides of the processor. This implementation of six HBM2s per processor is the world's first and was developed in collaboration with NEC, the Taiwan Semiconductor Manufacturing Company Ltd. (TSMC), and Broadcom using the TSMC chip on wafer on substrate (CoWoS) technology [6]. Due to such a cutting-edge technology, the package provides the world's highest memory bandwidth of 1.22 TB/s per processor.

The vector processor mainly consists of 8 vector cores, a 16 MB last level cache (LLC), a 2D mesh memory network, and six HBM2 memory interfaces. In VE Type 10A, each core operates at a frequency of 1.6 GHz, which achieves 614.4 Gflop/s for single-precision floating-point operations and 307.2 Gflop/s for double-precision floating-point operations. The 16 MB LLC is connected to each core with a 2D mesh. The bandwidth of LLC realizes 3.0 TB/s. Each HBM2 memory interface is connected to HBM2 at a bandwidth of 204.8 GB/s. The vector processor is connected to VH and DMA engines through the PCI express Gen3 interface. The DMA engine transfers data between VE and VH. The vector processor is manufactured by a 16 nm FINFET process, and about 4.8 billion transistors are integrated in an area of 14.96 × 33.00 mm. The power of VE and HBM2 memory is designed to 300 W or less for a single VE card.

2.2 Execution Model of SX-Aurora TSUBASA

In SX-Aurora TSUBASA, applications are executed on VE. VH is responsible for OS functions such as process scheduling and system call processing from applications. Fundamentally different from the execution model of accelerators such as GPU, the execution model of SX-Aurora TSUBASA can minimize data transfers between VE and VH since an application is executed on VE. Furthermore, as OS related functions such as system calls are automatically offloaded to VH, no special specification is not necessary for the offloading.

Moreover, as GNU C library (glibc) has been ported to VE, the standard programming environment can be used. In other words, no special programming is required to develop an application for SX-Aurora TSUBASA. A standard programming language such as C/C++ and Fortran can be used for SX-Aurora TSUBASA. This achieves the high usability as the previous vector supercomputers.

3 Evaluation

This section evaluates the basic performance of SX-Aurora TSUBASA using three benchmark programs; the Stream benchmark, the Himeno benchmark, and the HPCG benchmark.

3.1 Experimental Environments

For the evaluation, SX-Aurora TSUBASA A300-2 with one VE Type 10B was used. Table 1 shows the specifications of VE and VH of SX-Aurora TSUBASA used for the evaluation. Since the frequency of VE is 1.40 GHz, the single-precision and double-precision floating-point performances become 4.30 Tflop/s and 2.15 Tflop/s, respectively. The memory bandwidth is 1.22 TB/s and the cache bandwidth is 2.66 TB/s. For VH, Intel Xeon Gold 6126 processor is used. The single-precision and double-precision floating-point performances are 1996.8 Gflop/s and 998.4 Gflop/s, respectively. The memory bandwidth is 128 GB/s. For comparison, SX-ACE, Tesla V100, Xeon Phi 7290 are used as also shown in Table 1.

The Stream benchmark for measuring memory bandwidth performance [7], the Himeno benchmark for Jacobian calculation that frequently used in scientific and technical calculations [8], the HPCG benchmark developed as a benchmark close to the behavior of practical applications [9] are used.

3.2 Performance Evaluation of SX-Aurora TSUBASA

Figure 2 shows the sustained memory bandwidth measured by using the Stream benchmark. The horizontal axis shows the processors. The vertical axis shows the sustained memory bandwidth. Figure 2 shows that SX-Aurora TSUBASA achieves a high memory bandwidth performance of about 1.0 TB/s. On the other hand,

Table 1 Specifications of processors used in the evaluations

Processor	Vector engine type 10B	Intel Xeon Gold 6126	SX-ACE
Frequency	1.40 GHz	2.60 GHz	1.0 GHz
Performance/core	537.6 Gflop/s (SP) 268.8 Gflop/s (DP)	166.4 Gflop/s (SP) 83.2 Gflop/s (DP)	64.0 Gflop/s (SP/DP)
Number of cores	8	12	4
Performance/socket	4.30 Tflop/s (SP) 2.15 Tflop/s (DP)	1996.8 Gflop/s (SP) 998.4 Gflop/s (DP)	256 Gflop/s (SP/DP)
Memory subsystem	HBM2 × 6 modules	DDR4-2666 DIMM × 6 channels	DDR3-2000 DIMM × 16 channels
Memory bandwidth	1.22 TB/s	128 GB/s	256 GB/s
Memory capacity	48 GB	96 GB	64 GB
LLC bandwidth	2.66 TB/s	N/A	1.0 TB/s
LLC capacity	16 MB shared	19.25 MB shared	1 MB private

Processor	NVIDIA Tesla V100	Intel Xeon Phi 7290
Frequency	1.245 GHz	1.50 GHz
Performance/core	27.343 Gflop/s (SP) 13.671 Gflop/s (DP)	96 Gflop/s (SP) 48 Gflop/s (DP)
Number of cores	5120	72
Performance/socket	14 Tflop/s (SP) 7 Tflop/s (DP)	6.912 Tflop/s (SP) 3.456 Tflop/s (DP)
Memory subsystem	HBM2	MCDRAM + DDR4-2400 × 6 channels
Memory bandwidth	900 GB/s	N/A (MCDRAM), 115.2 GB/s (DDR)
Memory capacity	16 GB	16 GB (MCDRAM), 96 GB (DDR)
LLC bandwidth	N/A	N/A
LLC capacity	L2 6 MB shared, L1 128 KB private	1 MB shared by two cores

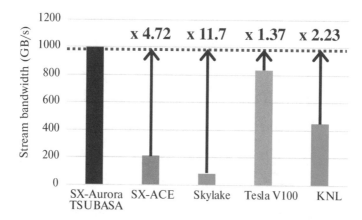

Fig. 2 Performance of the Stream benchmark

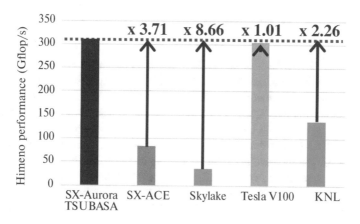

Fig. 3 Performance of the Himeno benchmark

SX-ACE, Skylake, Tesla V100, and KNL have achieved memory bandwidths of 211 GB/s, 85 GB/s, 727 GB/s, and 446 GB/s, respectively. By using the six HBM2 integration technology, the sustained memory bandwidth of SX-Aurora TSUBASA is about 4.7, 11.4, 1.37, and 2.23 times higher than those of SX-ACE, Skylake, Tesla V100, and KNL, respectively. The efficiencies of SX-Aurora TSUBASA, SX-ACE, Skylake, and V100 for the theoretical memory bandwidth are 81.8%, 83%, and 66%, and 81%, respectively. It is shown that SX-Aurora TSUBASA and SX-ACE, which are vector type supercomputers, achieve higher efficiency than Skylake. This is because vector supercomputers are designed with emphasis on memory performance.

Figure 3 shows the performance of the Himeno benchmark. The horizontal axis shows the processors. The vertical axis shows the performance of the Himeno benchmark. Figure 3 shows that the performance of SX-Aurora TSUBASA is higher than that of SX-ACE, Skylake, Tesla V100, and KNL. The Himeno performance

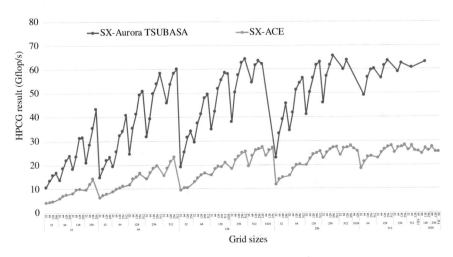

Fig. 4 Performance of the HPCG benchmark

of SX-Aurora TSUBASA is about 3.71, 8.66, 1.01, and 2.26 times higher than those of SX-ACE, Skylake, Tesla V100 and KNL, respectively. This is because the high memory bandwidth of SX-Aurora TSUBASA contributes to the high sustained performance. The Himeno benchmark is well-known as a benchmark that requires high memory bandwidth. Thus, the highest memory bandwidth of SX-Aurora TSUBASA contributes to the higher performance than other processors.

Figure 4 shows the results of the HPCG benchmark. The horizontal axis shows the grid size that can be configured when executing the HPCG benchmark. The vertical axis shows the performance of the HPCG benchmark. This result shows that the performance of SX-Aurora TSUBASA is higher than that of SX-ACE. Comparing the highest performance of SX-Aurora TSUBASA and SX-ACE, SX-Aurora TSUBASA achieves about 2.34 times higher performance than SX-ACE at maximum. Since the HPCG benchmark is also known to be a benchmark that requires memory bandwidth, the high memory bandwidth of SX-Aurora TSUBASA has achieved high performance.

4 Conclusions

This paper evaluates the basic performance of the latest vector supercomputer SX-Aurora TSUBASA.

First, by measuring the sustained memory performance of SX-Aurora TSUBASA using the Stream benchmark, it is clarified that a high sustained memory bandwidth can be achieved. Then, the performances of SX-Aurora TSUBASA were investigated using more practical benchmarks such as the Himeno benchmark and HPCG benchmark. As a result, SX-Aurora TSUBASA can achieve higher sustained performance than SX-ACE and other scalar processors due to the its high memory bandwidth.

For future work, more performance and power evaluations using real applications need to be conducted to further clarify the potential of SX-Aurora TSUBASA.

Acknowledgements This research was supported in part by MEXT as "Next Generation High-Performance Computing Infrastructures and Applications R&D Program," entitled "R&D of A Quantum-Annealing-Assisted Next Generation HPC Infrastructure and its Applications."

References

1. TOP500 supercomputer sites. http://www.top500.org/
2. Soga, T., Musa, A., Shimomura, Y., Egawa, R., Itakura, K.I., Takizawa, H., Okabe, K., Kobayashi, H.: Performance evaluation of NEC SX-9 using real science and engineering applications. In: Proceedings of the Conference on High Performance Computing Networking, Storage and Analysis, ser. SC '09, pp. 28:1–28:12 (2009). http://doi.acm.org/10.1145/1654059.1654088
3. Egawa, R., Komatsu, K., Momose, S., Isobe, Y., Musa, A., Takizawa, H., Kobayashi, H.: Potential of a modern vector supercomputer for practical applications: performance evaluation of SX-ACE. J. Supercomput. **73**(9), 3948–3976 (2017). https://doi.org/10.1007/s11227-017-1993-y
4. Komatsu, K., Egawa, R., Isobe, Y., Ogata, R., Takizawa, H., Kobayashi, H.: An approach to the highest efficiency of the HPCG benchmark on the SX-ACE supercomputer. In: Proceedings of the Conference on High Performance Computing Networking, Storage and Analysis (SC15), Poster, pp. 1–2 (2015)
5. Dongarra, J., Heroux, M.A., Luszczek, P.: High-performance conjugate-gradient benchmark. Int. J. High Perform. Comput. Appl. **30**(1), 3–10 (2016). https://doi.org/10.1177/1094342015593158
6. Hou, S.Y., Chen, W.C., Hu, C., Chiu, C., Ting, K.C., Lin, T.S., Wei, W.H., Chiou, W.C., Lin, V.J., Chang, V.C., Wang, C.T., Wu, C.H., Yu, D.: Wafer-level integration of an advanced logic-memory system through the second-generation CoWoS technology. IEEE Trans. Electron Devices **64**(10), 4071–4077 (2017)
7. McCalpin, J.D.: Memory bandwidth and machine balance in current high performance computers. In: IEEE Computer Society Technical Committee on Computer Architecture (TCCA) Newsletter, pp. 19–25 (1995)
8. Himeno Benchmark. http://accc.riken.jp/en/supercom/himenobmt/
9. HPCG Benchmark. http://www.hpcg-benchmark.org/

Optimized COAWST for the Vector Supercomputer SX-ACE

Shivanshu Kumar Singh, Kota Sakakura, Sourav Saha, Koji Goto, Raghunandan Mathur, Osamu Watanabe, and Akihiro Musa

Abstract Tropical cyclones cause immense damage with destructive winds, storm surges, and heavy rainfall with flooding in coastal regions. The intensity and frequency of tropical cyclones are expected to increase as a result of climate change. Therefore, the impact of high-intensity tropical cyclones, i.e., supertyphoons under the climate change needs to be researched. Our goal is to study the characteristics of supertyphoons under different conditions using a scientific application, named Coupled Ocean-Atmosphere-Wave-Sediment Transport (COAWST), and to optimize COAWST to predict the probable damage with proper warning and adequate accuracy on the NEC SX-ACE vector supercomputer. COAWST was developed by the US Geological Survey (USGS) to understand coastal changes caused by natural processes, and SX-ACE is a modern supercomputers with powerful vector cores, and is widely used to solve the large scale issues related to climatology and meteorology. In this paper, we proposed some vectorization strategies on SX-ACE that improve the computational performance of COAWST. Our proposed vectorization strategies have improved performance of COAWST up to 62.7% as compared to its original version for simulation. This paper aims to showcase the importance of the vectorization technology in order to speedily and accurately simulate supertyphoons related to coastal disasters.

S. K. Singh (✉) · S. Saha · R. Mathur
NEC Technologies India, Noida, U.P., India
e-mail: shivanshu.singh@india.nec.com; sourav.saha@india.nec.com;
raghunandan.mathur@india.nec.com

K. Sakakura · K. Goto · O. Watanabe · A. Musa
NEC Corporation, Tokyo, Japan
e-mail: k-sakakura@ay.jp.nec.com; k-goto@cq.jp.nec.com; o-watanabe@az.jp.nec.com;
a-musa@bq.jp.nec.com

© Springer Nature Switzerland AG 2020
M. M. Resch et al. (eds.), *Sustained Simulation Performance 2018 and 2019*,
https://doi.org/10.1007/978-3-030-39181-2_8

1   Introduction

Tropical cyclones cause immense damage with destructive winds, storm surges, heavy rainfall and flooding in coastal regions. The high intensity tropical cyclones i.e. supertyphoons are expected to increase as a result of climate change [1–3]. Supertyphoons affect people and infrastructure in coastal regions. With rapid urbanization of coastal regions, it is very important to research for faster simulation of such supertyphoons in order to help the governments and local administrative bodies of at-risk areas to mitigate the risk of damage and take preventive measures to minimize the severe impact in coastal regions.

The supertyphoon Haiyan, one of the strongest supertyphoons recorded in last few decades, caused enormous damage to the coastal regions of the Philippines, Vietnam, and nearby areas in November 2013 [2]. its minimum central pressure was 895 hPa, and maximum peak gust speed was over 90 m/s [2]. According to the final report of the National Disaster Risk Reduction and Management Council 2013, this enormous supertyphoon resulted in 6300 fatalities, 28,688 injured and 1062 missing [2].

The faster and more accurate computer simulations of such supertyphoons are a big challenge because they require high-performance computational power. Therefore, in our research, we use a NEC SX-ACE vector supercomputer (hereafter SX-ACE) to perform a high-speed and accurate numerical simulation of a Haiyan-like supertyphoon case using the Coupled Ocean-Atmosphere-Wave-Sediment Transport (COAWST) modeling system. COAWST is an open source scientific application designed for understanding coastal changes caused by natural processes.

Typhoon Haiyan continued for 5 days. In our earlier research [4], we attempted to perform a computer simulation for 3 days of Typhoon Haiyan in 24 computational hours using COAWST on the earth simulator. We achieved this with 38% faster COAWST compared with original performance. In this research paper, we attempt to complete a simulation of 5-day Haiyan test case with two-level nested domains of the computational grid size of $(1334 \times 667 \times 56)$ and $(2002 \times 703 \times 56)$ to be executed within 24 computational hours using vectorization of COAWST on SX-ACE.

2   Characteristics of SX-ACE

SX-ACE is a NEC SX series vector supercomputer. Its processor is comprised of four cores, which can provide a double-precision floating point operating rate of 256 Gflop/s with a memory bandwidth of 256 GByte/s, and a memory capacity of 64 GByte [5]. Each SX-ACE core is composed of three major components: a Scalar Processing Unit (SPU), a Vector Processing Unit (VPU), and a large capacity Assignable Data Buffer (ADB) [6, 7]. Each core is connected to Memory Control Unit (MCU) at a bandwidth of 256 GB/s through the memory crossbar, and four cores of one processor share the memory bandwidth, as shown in Fig. 1.

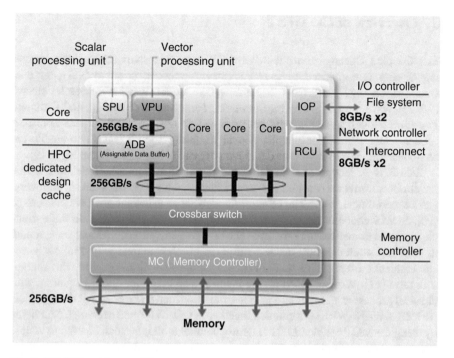

Fig. 1 SX-ACE architecture [8]

The VPU is a major component of the SX-ACE core that is connected with the vector cache ADB. It can process up to 256 vector elements (vector length of VPU) with a single vector instruction. The ADB is implemented to avoid the frequent memory access of vector load operation by retaining the reusable data on a chip.

The functionality of SPU is very limited in SX-ACE. It decodes all instructions, processes the scalar instructions, and transfers all vector instructions to the VPU, which processes the vector instructions.

The FORTRAN90/SX is a Fortran compiler with language specifications that conform to the international standard (ISO/IEC 1539-1:1997). It provides advanced automatic vectorization and optimization functions. The MPI/SX is the implementation of message-passing interface (MPI) version 3.0, and is available for parallel programming, which uses shared memory functions for communications within a node, and directly uses the inter node crossbar switch (IXS) functions for communications between nodes to achieve high-performance communication.

3   Overview of COAWST

The Coupled Ocean-Atmosphere-Wave-Sediment Transport (COAWST) Modeling System is composed of four components models: ocean model (Regional Ocean Modelling System (ROMS)), atmosphere model (Weather Research Forecast (WRF)), wave model (Sea Wave Simulating Near Shores (SWAN)), and Sediment Transport Model. In our current research, we do not use the Sediment Transport Model. COAWST uses a coupler Model Coupling Toolkit (MCT) to enable the component models to exchange data field with each other as shown in Fig. 2 [9].

ROMS is an incompressible, hydrostatic, primitive equation model with a free sea surface, horizontal curvilinear coordinates, and a generalized terrain-following vertical coordinate, which can be configured to enhance resolution at the sea surface [10]. ROMS contains various preprocessing options to enable/disable the various physical and numerical options. SWAN is a third-generation numerical wave model to compute random, short-crested waves in coastal regions with shallow water and ambient currents. This model is based on the advection equation with source-sink term [11]. WRF is based on the incompressible Navier–Stokes equation with physical processes: cloud micro physics, radiation, and so on [12]. All three models (ROMS, SWAN, WRF) are coupled together in COAWST and exchange data fields by using the MCT coupler. MCT is a fully parallel toolkit and can be used to couple message-passing parallel models in order to create a parallelized coupled model [13].

Table 1 shows the vector performance of ROMS, SWAN, and WRF in the original COAWST. ROMS has a vector operation ratio of 98.00% and average vector length of 186.0. SWAN and WRF have vector operation ratio of 86.00 and 92.00% with shorter vector length of 132.0 and 90.0, respectively. Therefore, ROMS is highly vector friendly but SWAN and WRF are not. So, WRF and SWAN cannot achieve high performance on SX-ACE. We propose some vectorization strategies on SX-ACE to reduce the computation time of the individual models in COAWST.

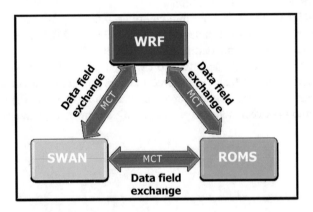

Fig. 2 Exchange of data fields between each model

Table 1 Vector characteristics of original COAWST

Model name	Vector operation ratio	Average vector length
ROMS	98.00%	186.0
SWAN	86.00%	132.0
WRF	92.00%	90.0

4 Optimization Strategies

As mentioned in Sect. 2, VPU is the most important component of SX-ACE. Thus, the performance of an application highly depends on its vector operation ratio and average vector length. In our earlier research paper [4], we proposed four vectorization strategies to improve the vector operation ratio and average vector length of subroutines. In addition, we propose some more optimization strategies in this paper.

4.1 To Increase Vector Operations

The high number of vector operations in a vectorized do-loop will result in better performance on the vector processor of SX-ACE. In this section, we discuss some vectorization strategies to increase the vector operations.

1. *Removal of RAW Data Dependencies*
   The Read-After-Write (RAW) data dependency inhibits the vectorization of do-loops. This issue occurs when a statement in a do-loop accesses a data element of array in ith iteration and accesses the same data element in next iterations of the same do-loop. These data dependencies can be resolved by restructuring the do-loop. If a RAW dependency is unknown to the compiler, then we can analyze the loop structure and instruct the compiler with a compiler directive (e.g., *"!CDIR NODEP"*) to resolve such data dependencies.
2. *Removal of I/O Statements from Computational Do-Loops*
   The I/O statements are important in programming language for transferring data to or from I/O devices but the I/O statement inside a do-loop is a vector inhibitor. Therefore, the I/O statement inside a computation intensive vectorized do-loop affects its performance on a vector processor. In such cases, we divide do-loop into two parts to separate the I/O intensive code and computation intensive code of the original do-loop. The compiler can vectorize the computation intensive code. In this way, we can improve the overall performance of the do-loop.
3. *Removal of Subroutine Calls from Computational Do-Loops*
   The user defined subroutine and unvectorizable intrinsic function call inside a do-loop are vectorization inhibitor. The modern compilers can perform the inline expansion of the user defined subroutine for vectorization. However, if a callee subroutine contains attributes like NAMELIST, DATA, and SAVE etc., then the compiler cannot perform the inline expansion of the callee in the caller do-loop

for vectorization. In the case of variable with a DATA attribute, we can move the variable from the callee subroutine to the caller subroutine and pass the variable as an argument. In other cases, if the inline expanded image of the callee subroutine in a do-loop contains the allocate/deallocate statements, then the compiler cannot vectorize such a loop even after the inline expansion of the callee subroutines. In this case also, we can allocate the array from the callee subroutine into the caller subroutine and pass the array as argument to the callee subroutine. These type of manual code modifications can vectorize the do-loops with subroutine and improve the vector operations in application.

4.2   To Increase Loop Length

As one vector core of SX-ACE processor can process up to 256 elements in one operation, sufficient loop length of vectorized code increases the effect of vectorization. In this section, we explain some optimization strategies to increase the loop length of a nested do-loop structure.

1. *Loop-Interchange*
   If a vectorized two level nested do-loop performs calculation along the small dimensions of an array, then a loop-interchange can be done to perform calculation along the large dimensions and improve the loop length of vectorized do-loop. In the case of a three-dimensional array, when a subroutine performs calculation on small dimension and subroutine calls is implemented inside a two-level nested do-loop on large dimensions, then loop interchange is not a simple optimization strategy. In this case, we can move the do-loop on large dimension of array from the caller subroutine to the callee subroutine and the calculation on large dimension in the callee subroutine is vectorized. In this way, we can increase the vector length of a vectorized calculation loop in the callee subroutine.

2. *Loop-Collapse*
   The loop-collapse is an important optimization strategy to increase the loop length of small length nested do-loop structure by collapsing nested do-loop in to a single level do-loop. The compilers can perform the loop-collapse automatically by judging the loop length of a nested do-loop. However, when indices of a nested do-loop are stored in a work-array, then the compiler cannot judge the loop indices. In this case, code needs to be manually modified for loop collapse to increase the vector length of the loop structure.

4.3 To Optimize IF-ELSE Branch in a Do-Loop

An IF-ELSE statement in a vectorized nested do-loop structure degrades the performance of the do-loop. If the condition of the IF-ELSE statement does not depend on the loop variable, we can split the computation of complete do-loop in two separate code sections on the basis of the IF-ELSE condition. After splitting the computation, we can keep the corresponding code section in the IF and ELSE section. In this way, we can optimize the computation of a do-loop.

4.4 To Optimize the Memory Allocation

The allocation and deallocation of data structure includes keeping track of the state of allocated blocks, searching for a free memory block, fragmentation of memory block etc. These hidden tasks affect the performance of a processor. Thus, the dynamic memory allocation is a high-cost statement. The large number of allocations and deallocations reduces the performance of an application. Therefore, we can re-use the already allocated memory block at different stages of calculation and reduce the overhead of allocation and deallocation.

5 Optimization Examples

In this section, we discuss the implementation of optimization strategies. The implementations of optimization strategies (Sects. 4.1–4.3) are described in our earlier research paper [4]. In this research paper, we discuss the implementation of optimization strategy (Sect. 4.4).

5.1 To Optimize Memory Allocation

The WRF model uses an intermediate domain that has a resolution equal to the inner domain for the data exchange as shown in Fig. 3. The WRF model performs the data exchange from the outer domain to the inner domain and vice versa after every timestep via the intermediate domain as shown in data exchange order ①, ②, ③, and ④ in Fig. 3. WRF allocates and deallocates intermediate domains and their member data structure before and after every data exchange. At each data exchange, WRF allocates an approximately 1200-member data structure of the intermediate, which is a very high-cost operation. To eliminate this overhead, we can allocate the intermediate domain at the first data exchange and keep it allocated till the end of the simulation. In this way, we can reduce the overhead of the number of allocations

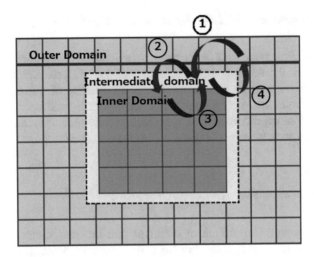

Fig. 3 Domain orientation and two-way data exchange between nested domains in WRF. The number ①, ②, ③, and ④ show the data exchange order in WRF. The data exchange order ① and ② happens after calculation in the outer domain and ③ and ④ happens after calculation in the inner domain

and deallocations from ($1200 \times n$ timesteps) to (1200×1 timestep), which is large performance improvement. Here, n denotes the number of timesteps.

6 Performance Evaluation

In this section, we discuss the performance evaluation of the original and optimized COAWSTs on the Haiyan dataset.

6.1 Description of Haiyan Data

Our Haiyan dataset has one domain each in ROMS, and SWAN and two domains in WRF. The outer domains of ROMS and SWAN have the same resolution as the outer domain of WRF. The grid spacing is 3 km for the outer domain and 1 km for the inner domain in WRF. The simulation domain covers 109.2° East to 150.75° East and Equator to 20° North. The numbers of grid points in the domains of ROMS, SWAN, and WRF are shown in Table 2.

On the basis of the computation ratio of each model in COAWST, we evaluated the performance of the original and optimized versions of COAWST using 353 nodes of SX-ACE: 3 nodes (12 MPI processes) for ROMS, 100 nodes (400 MPI processes) for SWAN, and 250 nodes (1000 MPI processes) for WRF for a

Table 2 Computational domain size of Haiyan dataset

Model	Domain no.	Domain size in X-direction	Domain size in Y-direction	Domain size in Z-direction
ROMS	Domain 1	1334	667	40
SWAN	Domain 1	1334	667	–
WRF	Domain 1	1334	667	56
	Domain 2	2002	703	56

simulation time of 3 days on the Haiyan dataset. We used an SX-ACE profiler *ftrace* to measure the evaluation parameter of the original and optimized COAWSTs.

6.2 Original and Optimized Subroutines

We have examined two parameters (the vector operation ratio (Vec. ratio) and the average vector length (Avg. vec. length)) using SX-ACE profiler to evaluate the performance of high-cost subroutines before and after optimization on SX-ACE for a small simulation time of 6 h, as shown in Table 3. As calculated in Table 3, we have achieved the Vec. ratio of more than 97.0% for 13 out of 18 optimized subroutines. As a result of optimization, the Avg. vec. length is increased to more than 100 for 14 out of 16 optimized subroutines, out of which Avg. vec. length of 7 subroutines is more than 200. This improvement in vector performance reduced the execution time for 12 out of 18 optimized subroutines by over 75.0% of the original execution times, as shown in Fig. 4.

Subroutine *gls_corstep_tile* of ROMS, *swsnl3*, *adddis*, *swpsel* of SWAN, and *clphyld*, *advect_scalar_pd* of WRF were very calculation intensive and high-cost in the original COAWST. The 3.9% improvement in the Vec. ratio of the optimized *gls_corstep_tile* reduced the execution time by 65.8% of the original *gls_corstep_tile*. The execution time of the highest cost subroutine *swsnl3* of SWAN is reduced by 89.7% with respect to the original *swsnl3* because of improvement in vec. ratio and Avg. vec. length. The optimized *clphyld* has a Vec. ratio of 98.1% and Avg. vec. length of 98.0. Despite the small improvement in Avg. vec. length in the original *clphyld*, the execution time of the optimized *clphyld* was reduced by 80.0% of the original *clphyld* because of the increase in Vec. ratio. The execution time of optimized *advect_scalar_pd* was reduced by 87.2% of the original *advect_scalar_pd* due to the improvement in its Vec. ratio. The Vec. ratio and Avg. vec. length of the original subroutine *strsd*, *strsxy*, *w_damp* were all 0.00% and 0.0 respectively. The performance of these optimized subroutine is improved by over 84.0% as a result of improvement in the Vec. ratio and their Avg. vec. length is improved to over 98.0%. These performance statistics confirm that our vectorization-specific optimization strategies are suitable to improve the executional performance of high-cost subroutines in COAWST on SX-ACE.

Table 3 Improvement in vectorization metrics with optimization strategy using Haiyan data for a simulation time of 6 h

Model	Subroutines	Original			Optimized		
		Exec. time (s)	Vec. ratio (%)	Avg. vec. length	Exec. time (s)	Vec. ratio (%)	Avg. vec. length
ROMS	gls_corstep_tile	954.943	95.17	168.3	327.027	99.08	168.3
SWAN	swsnl3	613.958	31.39	45.3	63.074	97.99	120.3
	adddis	528.187	21.83	8.1	105.038	97.63	212.0
	swpsel	358.681	4.21	47.4	163.349	95.51	145.2
	strsd	218.984	0.00	0.0	33.053	99.77	225.9
	strsxy	204.574	0.00	0.0	32.801	99.56	228.4
	sintgrl	173.966	93.54	112.4	76.052	94.98	95.6
	sprosd	133.869	2.22	36.0	27.094	98.92	154.6
	swcap	117.148	9.94	31.4	67.877	91.85	193.2
	ssurf	100.183	76.19	45.2	22.721	99.23	224.4
	rescale	96.947	97.01	127.2	50.925	98.82	193.8
	filnl3	92.684	75.76	51.6	18.877	99.61	227.2
	swind3	84.419	4.33	25.0	19.693	99.54	225.4
	philim	68.370	95.86	107.0	24.174	99.41	219.4
WRF	clphy1d	464.231	62.64	91.0	92.618	98.11	98.0
	advect_scalar_pd	316.012	78.41	81.5	40.315	98.74	79.6
	w_damp	192.939	0.00	0.0	21.949	98.40	155.2
	in_use_for_config	1242.136	0.00	0.0	0.928	0.00	0.0

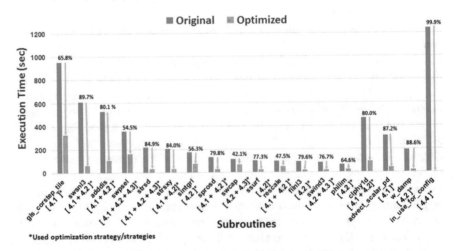

Fig. 4 Performance improvement of optimized subroutines in COAWST using Haiyan data for simulation time of 6 h

6.3 Original and Optimized COAWST

The overall execution time of the COAWST involves some inter-model waiting time for data exchange after a fixed interval of simulation time because of different computational loads of ROMS, SWAN, and WRF. Thus, to demonstrate the actual computational gain in the COAWST, only computation time of each model is considered by subtracting their individual waiting times from their individual execution times. From the performance results shown in Fig. 5, the execution time of ROMS, SWAN, and WRF are reduced by 22.13%, 56.53%, and 67.3%, respectively, in comparison to their respective original versions on SX-ACE for a simulation time of 6 h using the Haiyan data. As we aimed to complete a simulation of 5 days within 24 h computation time, the optimized COAWST completed the simulation of 5 days in 21.69 h computation time, whereas the original COAWST completed the same simulation in 34.96 h on SX-ACE. Hence, we have achieved aim to optimize the COAWST to complete the simulation of 5 days of the Haiyan dataset within 24 h computation time on the vector supercomputer SX-ACE.

7 Summary

Tropical cyclones cause immense damages with destructive winds, storm surges, and heavy rainfall with flooding in coastal regions. The intensity and frequency of tropical cyclones is expected to increase due to climate change. The number of high-intensity typhoons, i.e., supertyphoons, is expected to increase. Therefore, such supertyphoons must be researched, and a faster model for their simulations must be developed. In our research, we used an open source coupled application named Coupled Ocean-Atmosphere-Wave-Sediment Transport (COAWST) that is composed of Regional Ocean Modelling System (ROMS) as an ocean model, Sea Wave Simulating Near Shores (SWAN) as a wave model, and Weather Research Forecast (WRF) as an atmospheric model and couples them by using a Model

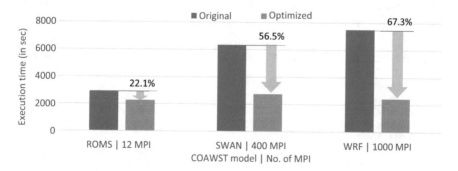

Fig. 5 COAWST performance improvement using Haiyan data for a simulation time of 6 h

Coupler Toolkit (MCT). In our earlier research [4], we optimized the COAWST model by 38% to its original COAWST for a simulation on the Haiyan dataset. In a continuation of our earlier research, we further optimized COAWST by 62.7% of its original performance for a simulation of 5 days on the Typhoon Haiyan dataset. Our current optimized COAWST is faster than the COAWST version presented previously [4].

Our research continues to improve the computation time of COAWST with a vectorization strategy and evaluation of the optimized COAWST on other single instruction, multiple data (SIMD) architectures.

Acknowledgements The authors thanks Dr. Nobuhito Mori and Dr. Tetsuya Takemi of Disaster Prevention Research Institute, Kyoto University, Japan, and Dr. Junichi Ninomiya of Kanazawa University, Japan, for providing the Haiyan test data and guidance in verifying results of COAWST. This research is supported by the Environment Research and Technology Development Fund (2-1712) of the Ministry of the Environment, Japan.

References

1. Mori, N., Takemi, T.: Impact assessment of coastal hazards due to future changes of tropical cyclones in the North Pacific Ocean. Weather Climate Extremes **11**, 53–69 (2016). ISSN 2212-0947. https://doi.org/10.1016/j.wace.2015.09.002
2. Mori, N., Kato, M., Kim, S., Mase, H., Shibutani, Y., Takemi, T., Tsuboki, K., Yasuda, T.: Local amplification of storm surge by Super Typhoon Haiyan in Leyte Gulf. Geophys. Res. Lett. **41**, 5106–5113 (2014). https://doi.org/10.1002/2014GL060689
3. Tsuboki, K., Yoshioka, M.K., Shinoda, T., Kato, M., Kanada, S., Kitoh, A.: Future increase of supertyphoon intensity associated with climate change. Geophys. Res. Lett. **42**, 646–652 (2015). https://doi.org/10.1002/2014GL061793
4. Kumar Singh, S., et al.: Optimizations of COAWST for a large simulation on the earth simulator. In: 2018 IEEE International Conference on Cluster Computing (CLUSTER), Belfast, 2018, pp. 629–636. https://doi.org/10.1109/CLUSTER.2018.00080
5. Egawa, R., Komatsu, K., Momose, S., Isobe, Y., Musa, A., Takizawa, H., Kobayashi, H.: Potential of a modern vector supercomputer for practical applications: performance evaluation of SX-ACE. J. Supercomput. **73**(9), 3948–3976 (2017). https://doi.org/10.1007/s11227-017-1993-y
6. Momose, S., Hagiwara, T., Isobe, Y., Takahara, H.: The brand-new vector supercomputer, SX-ACE. In: Kunkel, J.M., Ludwig, T., Meuer, H.W. (eds.) Supercomputing. ISC 2014. Lecture Notes in Computer Science, vol. 8488. Springer, Cham (2014)
7. Momose, S.: NEC vector supercomputer: its present and future. In: Resch, M., Bez, W., Focht, E., Kobayashi, H., Qi, J., Roller, S. (eds.) Sustained Simulation Performance 2015. Springer, Cham (2015)
8. Vector Supercomputer SX Series SX-ACE: http://de.nec.com/de_DE/en/documents/SX-ACE-brochure.pdf
9. Warner, J.C., Armstrong, B., He, R., Zambon, J.B.: Development of a coupled ocean-atmosphere-wave-sediment transport (COAWST) modeling system. Ocean Model. **35**(3), 230–244 (2010)
10. Marchesiello, P., McWilliams, J.C., Shchepetkin, A.: Open boundary conditions for long-term integration of regional oceanic models. Ocean Model. **3**(1–2), 1–20 (2001). ISSN 1463-5003. https://doi.org/10.1016/S1463-5003(00)00013-5

11. Zijlema, M.: Parallelization of a nearshore wind wave model for distributed memory architectures. In: Parallel Computational Fluid Dynamics 2004: Multidisciplinary Applications, pp. 207–214 (2005). https://doi.org/10.1016/B978-044452024-1/50027-0
12. Skamarock, W.C., Klemp, J.B., Dudhia, J., Gill, D.O., Barker, D.M., Duda, M.G., Huang, X.-Y., Wang, W., Powers, J.G.: 2008: A description of the advanced research WRF version 3. NCAR Tech. Note NCAR/TN-475+STR, p. 113. https://doi.org/10.5065/D68S4MVH
13. Larson, J., Jacob, R., Ong, E.: The model coupling toolkit: a New Fortran90 toolkit for building multiphysics parallel coupled models. Int. J. High Perform. Comput. Appl. **19**(3), 277–292 (2005). https://doi.org/10.1177/1094342005056115

Part III
Techniques and Tools for New-Generation Computing Systems

VEO and PyVEO: Vector Engine Offloading for the NEC SX-Aurora Tsubasa

Erich Focht

Abstract The SX-Aurora Tsubasa Vector Engine (VE) is NEC's latest instantiation of their long vector architecture for high performance computing and AI with large HBM2 memory of 48 GB and high memory bandwidth of 1.2 TB/s. It is completely different from the previous mainframe-sized product generations and comes as a PCIe card pluggable into normal Linux servers, where the VE integrates seamlessly into the Linux environment and runs native VE programs compiled with C, C++ or Fortran transparently, from the command line. This report introduces Vector Engine Offloading, VEO, the base mechanisms used to extend the programming model of the VE to an accelerator style offloaded model somewhat similar to OpenCL or CUDA. This programming model extends the scope of the VE and simplifies the porting of applications which have already been adapted to using accelerators like GPGPUs. The PyVEO Python bindings furthermore simplify accessing the power of the VE even from scripts and interactive notebooks.

1 Introduction

NEC has reimplemented its traditional SX long vector high bandwidth architecture into a completely new form, the SX-Aurora Tsubasa Vector Engine (VE). Instead of mainframe-sized cabinets with proprietary and custom made hardware, storage, operating system, interconnect, one can now build a vector machine with PCIe accelerator cards plugged into off-the-shelve Linux servers connected with Infiniband interconnect. The development of system software and tools has been opened up and the low entry barrier for the vector machines promises a much larger set of users and use cases for the machines.

The NEC SX-Aurora vector engine (VE) is a long vector processor which combines SIMD and pipelining. It has up to 48 GB 3D stacked HBM2 memory

E. Focht (✉)
NEC Deutschland GmbH, Stuttgart, Germany
e-mail: erich.focht@emea.nec.com

accessible with very high bandwidth of 1.22 TB/s and is packaged as a PCIe card pluggable into Linux servers. Each of the eight cores has a vector processing unit featuring 64 vector registers of the length of 256 * 64 = 16,384 bits, three vector fused multiply-add (FMA) units, two integer ALUs and a SQRT/DIV vector pipeline. The clock frequency of the VE is either 1400 or 1600 MHz, the higher frequency variant being water cooled. Each core delivers up to 307 GFLOPS (double precision)/614 GFLOPS (single precision, packed) and a memory bandwidth to the shared 16 MB last level cache of 409 GB/s [1, 2].

While the VE looks like a classical accelerator, it offers a wider range of programming models:

1. Native VE programming in C, C++, Fortran. Programs are running purely on the VEs and can use OpenMP and MPI with PeerDirect communication between VEs over PCIe or Infiniband for parallelization. Also they can call almost any Linux system call which is forwarded and executed on the vector host (VH) server.
2. Native VE programming with reverse offloading (VHcall). This is an extension of the native mode and allows VE programs to offload parts of the program to the VH, executing them on the x86_64 host system.
3. Main program running on VH with offloaded kernels running on the VE. This is the classical accelerator model as provided by CUDA, OpenCL and others. The VH program can use MPI and OpenMP, he offloaded kernels can use OpenMP and issue Linux system calls.
4. Hybrid MPI program with processes running on VH and VE, sharing one communicator.

This article focuses on the third programming model in the list above, its mechanisms and API.

With VEO NEC provides the base infrastructure for offloading parts of a VH program onto the VE. The API is somewhat similar to the 10 years old OpenCL (Open Computing Language), a framework for executing programs on heterogeneous hardware [3]. Like OpenCL, VEO explicitly sets arguments of called functions, the calls them asynchronously. But VEO does not contain mechanisms for just in time compilation of code and has a plain C API instead of being an extension of C++ requiring a separate compiler, like OpenCL.

Another widely used language for accelerator offloading is CUDA for nVIDIA GPGPUs [4]. It is an extension of the C++ language and requires a special compiler. The language uses special attributes to declare and address device memory and accelerator kernels and passes arguments to accelerator kernels similar to function call arguments. OpenCL and CUDA accelerator kernels are not supposed to use system calls and are not supposed to return any results, i.e. are of void type.

While OpenCL and CUDA require coding the heterogeneous part of the code rather explicitly, i.e. heavily changing the original source code, OpenMP [5] and OpenACC [6] take another approach: add compiler directives (pragmas in C/C++) which guide the compiler to "do the right thing", separate the accelerator code, create glue code that calls an accelerator specific runtime, create a "fat binary" which contains the host code and the accelerator code. These approaches simplify a lot the hybridization of existing applications. Of course, they require changing compilers and some underlying offloading framework they can build on. VEO is such a framework for the SX-Aurora Tsubasa VE.

The first VEO version was implemented as a C prototype in June 2015 by Erich Focht and tested on SX-Aurora Tsubasa simulators. After the vector engine silicon was available Teruyuki Imai re-implemented VEO in the currently available C++ and C form in October 2017. Since then VEO is being continuously developed and improved while its API is kept as stable as possible. The code is available on github [7] under the LGPL 2.1 license.

The next section of this article introduces the VE operating system VEOS and some peculiarities of running VE programs. It is followed by a description of VEO implementation details, then by a section describing the API of VEO. The PyVEO object oriented Python bindings to VE offloading follow with an architecture discussion and a short example. Finally the conclusion and outlook section lists a set of projects using VEO and provides an outlook to further developments.

2   Vector Engine OS

Initially, the primary intended usage mode for the SX-Aurora was the native VE mode for programs ported to run entirely on the VE. These programs are supposed to do IO, communicate and interact with the operating system, and "feel" as if they're running on a Linux system. With the difference that the VE has no Linux kernel running underneath, and actually has no kernel at all. VE register space is mapped onto the VH processors and accessible from kernel space, where it is controlled by the *ve_driver* kernel module. The VE operating system (VEOS) is a daemon running with root privileges which provides operating system services for a VE. Each VE in a system has its own instance of VEOS.

Figure 1 sketches the components involved in managing a VE and running user processes on it.

- The **VEOS** daemon is responsible for process management, scheduling, memory management and interfacing a user process to functionality provided by the ve_driver kernel module.
- **ived** is a daemon managing inter-VE communication resources.
- **vemmd** manages the mappings needed for VE communication over infiniband.
- **mmm** monitoring and maintenance manager brings up and controls the state of the VEs.
- The **pseudo-process** runs as a user process, manages the loading of the VE program into the accelerator and acts as an exception and syscall handler while the VE program is running.

With VEOS running on the VH decisions about the virtual memory layout of VE processes as well as scheduling are taken on the VH. VE processes can use 2 and

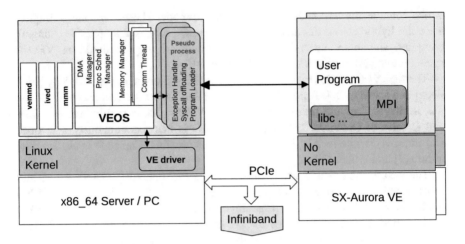

Fig. 1 VEOS components and VE process overview

64 MB page sizes for their virtual memory in multi-user and multi-processing mode, but are limited to real memory mode. Due to the very high memory bandwidth and internal parallelism the VE lacks precise interrupts, therefore a running process must have all pages in place as well as all virtual address translations set. Page faults or TLB misses can not be recovered on the VE.

A context switch on the VE implies copying the context over to the VH into the memory of the responsible VEOS daemon instance. Because the 64 long vector registers contribute to the context with 2 kB each, context switches are rather expensive and have large latency. For long running compute workloads with few system calls the VE operating system offloading bears the positive effect of keeping OS jitter away. On the other hand heavily multithreaded programming patterns with more threads than cores perform significantly slower than on a CPU with OS kernel under the hood.

The opposite of the native VE usage model is the accelerator model with truly hybrid VH-VE programs, where the main program runs on the VH along with multiple other VH threads and offloads one or more compute kernels to the VE (Fig. 2). Many programs have been ported to run this way since GPGPUs have become popular.

VEOS has low level mechanisms for helping a user process to interact with the VE, but there is no easy way to start VE processes and communicate with them, as an accelerator model would need. VEO provides the basic mechanisms for this.

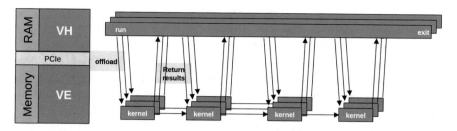

Fig. 2 VE as an accelerator: sketch of the hybrid VH-VE usage model with multiple VH threads, including the main thread, which offload VE kernel calls to multiple VE cores

3 Vector Engine Offloading Implementation

For an application programmer VEO consists of a library *libveo.so* which he links with the VH side program and which provides the API described in Sect. 4. Furthermore the user will need to split out the VE kernel functions, group and compile them into one or more dynamically shared objects for the VE.

Figure 3 is a rough sketch of the architecture of VEO. The VH program is creating a ProcHandle object which is used to control one vector engine. When the ProcHandle is created a *veorun helper* is loaded and started into the designated VE. This helper is providing the VE side mechanisms needed by VEO and runs as a program of the user owning the VH program. Internally and exposed as a C API to the user, ProcHandle has methods to load shared objects into the veorun helper process, allocate and free memory on the VE and start/stop ThreadContexts on the VH.

A ThreadContext object represents an additional thread on the VH which is associated to a worker thread on the VE. When a ThreadContext is instantiated the veorun helper is creating a worker thread on the VE by calling *clone()*, which

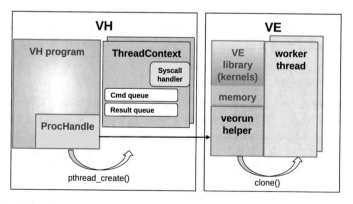

Fig. 3 The VEO architecture

is sharing its memory space. A ThreadContext has a command queue and a result queue. Asynchronous VE function calls are issued by enqueueing commands to the command queue. Once the corresponding worker thread is ready to execute a command, the parameters of the command are transfered to the VE, the function arguments and, if needed, the stack for the VE function are prepared, and finally the VE kernel function is invoked by the worker thread. After finishing execution the result is transfered into the result queue on the VH and the VH application can pick it up. Requests in a command queue are executed strictly in the submission order and will never be reordered.

ProcHandle methods are executed synchronously, e.g. allocating and freeing memory on the VE, loading dynamic shared objects or transferring data between memory buffers on VE and VH. ThreadContext methods are asynchronous: VH-VE memory transfers as well as VE kernel function calls.

4   Vector Engine Offloading API

This section describes the concrete mechanisms that a VEO program can use. First, it requires an include file:

```
#include <ve_offload.h>
```

The version of the VEO API is an integer and can be queried as follows:

```
int version = veo_api_version();
```

At the time of writing this article the API version is 4 and the VEO package version is 2.1.1.

4.1   Proc Handle and Contexts

A ProcHandle object is referenced in the VEO C API by a pointer to an opaque structure which is created as follows:

```
struct veo_proc_handle *proc;
proc = veo_proc_create(nodeid);
```

The parameter nodeid specifies the VE card ID on which the VEO process (the veorun helper binary) will be started. Enumerating the VE nodes is not in the scope of VEO, this can be done with the *libveosinfo* library.

An own veorun helper can be linked statically with the mk_veorun_static and replace the default binary. Its path can be either specified in the VEORUN_BIN environment variable or as second argument of the alternative proc creation call:

```
proc = veo_proc_create_static(nodeid, veorun_path);
```

After being finished with VEO the proc handle can be destroyed:

```
int rc = veo_proc_destroy(proc);
```

This operation is not mandatory because VEOS will clean up and kill the veorun helper process when the VH main process dies, anyway.

A ThreadContext object is also addressed in the C API by a pointer to an opaque structure. Contexts are created or opened inside procs by doing

```
struct *veo_thr_ctxt = veo_context_open(proc);
```

This creates a thread on the VH which controls the corresponding worker thread on the VE and becomes its "pseudo process" while VE kernels are executed. The VE worker thread is identified by the same TID as the VH thread. A user can chose to create multiple contexts, eg. one for each VE core, or just one context per VE. OpenMP VE kernels span multiple cores, therefore the interaction between worker threads and OpenMP threads should be carefully managed in order to avoid overcommitment of the VE cores.

A context can be closed by calling

```
int rc = veo_context_close(ctxt);
```

Contexts go through multiple states and under certain circumstances it can be useful to query them:

```
int res = veo_get_context_state(ctxt);
```

The result is one of

- `VEO_STATE_UNKNOWN`—the context has not yet been initialized, yet,
- `VEO_STATE_RUNNING`—the context is currently running a request,
- `VEO_STATE_SYSCALL`—the context is blocked for executing a syscall offloaded to the VH,
- `VEO_STATE_BLOCKED`—the context is currently stopped, with no request running,
- `VEO_STATE_EXIT`—the context's thread is exiting,

and should be interpreted with care due to the volatility of some of the states.

4.2 Libraries and Symbols

A shared library with VE kernels is loaded into the veorun process as follows:

```
uint64_t lib_h = veo_load_library(proc, lib_path);
```

The call returns a library handle that is needed for locating functions and symbols inside the library. The call is basically a `dlopen()` call on the VE side issued from inside the veorun helper.

Symbols inside the loaded libraries can be located as shown below:

```
uint64_t addr = veo_get_sym(proc, lib_h, sym_name);
```

The return value is the VE virtual address of the symbol inside the specific instance of the veorun helper. When using multiple proc instances, the addresses may be different for each of them.

When the veorun helper was linked statically, one can locate functions the same way as in a "normal" library but the `lib_h` argument must be set to zero.

4.3 Memory Buffer Allocation and Transfer

Allocating and freeing memory buffers are synchronous operations which require the proc handle as first argument. The memory is allocated in the address space of the veorun helper by calling `malloc()`, the buffer therefore lands on the VE processes heap.

```
uint64_t ve_addr;
int rc = veo_alloc_mem(proc, &ve_addr, len_bytes);
...
rc = veo_free_mem(proc, ve_addr);
```

In order to transfer data between some buffer on the VH and a memory buffer on the VE we can use two synchronous calls:

```
rc = veo_read_mem(proc, vh_buff, ve_addr, len);
```

and

```
rc = veo_write_mem(proc, ve_addr, vh_buff, len);
```

The function names are from the perspective of the VH program, i.e. `veo_read_mem()` transfers data from VE to VH while `veo_write_mem()` does it in the opposite direction. The asynchronous variants of these functions take as first argument the context and return a request ID:

```
req1 = veo_async_read_mem(ctxt,vh_buff,ve_addr,len);
req2 = veo_async_write_mem(ctxt,ve_addr,vh_buff,len);
```

These requests are queued into the context command queue the same way as asynchronous VE kernel calls and executed in the order determined by the enqueued requests although they are actually not VE kernels but VH functions.

Waiting for a request or querying its state is done the same way as described in Sect. 4.5.

4.4 VE Kernel Function Arguments

The arguments of an asynchronous VE function call are represented by a pointer to an opaque structure which is, respectively, allocated and freed by

```
struct veo_args *args = veo_args_alloc();
...
veo_args_free(arg);
```

The arguments object must be kept in memory until the result of the call has been collected. After the corresponding request was finished the arguments object can be reused after clearing it with

```
veo_args_clear(arg);
```

With one exception the asynchronous request arguments are prepared in a similar way to OpenCL: by setting each of them with a type specific function, for example:

```
rc = veo_args_set_u64(args, argnum, u64);
rc = veo_args_set_i64(args, argnum, i64);
rc = veo_args_set_float(args, argnum, float_f);
rc = veo_args_set_double(args, argnum, double_d);
```

argnum is the position of the argument in the called function and the last argument is its value. The functions ending with u64 and i64 are for 64 bit integer arguments and have corresponding equivalents for 32, 16 and 8 bit integers. The functions return zero when successful.

The Aurora Tsubasa VE ABI allows passing up to 8 arguments in registers. When a function has more than 8 arguments, they are all passed on the stack and the first 8 are also available in registers. Currently VEO is limited to passing up to 32 arguments to functions, but this limit might be increased in future.

The notable exception from OpenCL is the more complex function below:

```
rc = veo_args_set_stack(args,intent,argnum,buff,len);
```

It copies the content of a buffer *buff* with length *len* onto the VE caller stack, reserved for local variables of the caller function and passes the pointer to this copied buffer as the *argnum*th argument. The value of *intent* can be *VEO_INTENT_IN*, *VEO_INTENT_OUT* and *VEO_INTENT_INOUT*, the buffer will be accordingly copied in before the kernel execution, copied out after the kernel execution or copied in before and out after. This option can be used to easily pass in and out more complex arguments by reference, or to call Fortran functions which by default use "by reference" arguments passing.

4.5 Asynchronous VE Kernel Function Call

Finally, after all preparation, we can call the VE function asynchronously:

```
uint64_t req = veo_call_async(ctxt, addr, args);
```

The first argument is the context, the second is the address of the function in the context, i.e. the proc instance, as located with *veo_get_sym()*, and the third argument is the prepared arguments object, described in the previous section. The function returns a request ID which is incremented and starts with zero. In case of failure of the async request enqueueing the function returns *VEO_REQUEST_ID_INVALID*.

A more comfortable variant of the asynchronous function call does not require finding the function address before the call but takes the library handle *lhdl* and the function name as a string argument. The function's address on the VE is determined on the fly and cached in a hash:

```
req = veo_call_async_by_name(ctxt, lhdl, fname, args);
```

The status of a request can be queried non-blockingly:

```
int rc = veo_call_peek_result(ctxt, req, &result);
```

The result of this query is one of the values:

- *VEO_COMMAND_OK*—the function has finished normally,
- *VEO_COMMAND_EXCEPTION*—the function threw an exception on the VE,
- *VEO_COMMAND_ERROR*—execution error on VH side,
- *VEO_COMMAND_UNFINISHED*—the function did not finish, yet.

The *result* variable must be a 64 bit entity that will contain the value returned by the VE function. This is different from OpenCL and CUDA: VEO kernels can return values. Note that returning (Fortran) complex values is not supported by VEO, although defined in the ABI.

In order to block and wait for the request to finish, call

```
int rc = veo_call_wait_result(ctxt, req, &result);
```

5 PyVEO

This section describes an implementation of Python bindings to VEO which enable users to use VE offloading easily from inside Python programs. It extends the reach of VE accelerator programs to scripts and even interactive use. With PyVEO one can write interactive notebooks (e.g. Jupyter notebooks) with VE code that is interactively executed on a VE.

The Python module is available at github [8] and is implemented in Cython. It supports NumPy arrays and therefore VE acceleration can be integrated easily with packages used frequently in Python like SciPy.

PyVEO includes a class that facilitates building VE kernels from inside Python, thus has slightly extended functionality compared to VEO.

5.1 VE Offloading Objects and Methods

The Python bindings for VEO are not a one-to-one mapping of the C-API. Instead they are embedded into a Python object hierarchy that reflects to some extent the dependencies between the entities, as depicted in Fig. 4. IDs and VE addresses are wrapped into objects with appropriate methods.

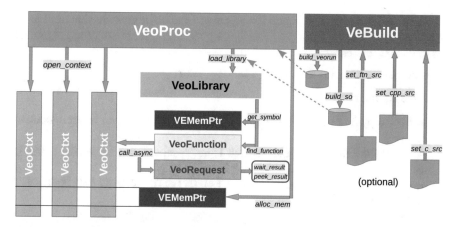

Fig. 4 Architecture of PyVEO. The VeBuild class is used for creating VE shared objects or statically linked veorun helpers including offload kernels. VeoProc is the initial class for VE offloading support

The *VeoProc* object corresponds to one running instance of the *veorun* helper program that controls one address space on a VE and enables its user to offload VE kernels. This object wraps the proc handle described in Sect. 4.1.

A *VeoLibrary* object represents a loaded dynamically shared object inside a *VeoProc* and is created by invoking its *load_library()* method.

A *VeoFunction* object represents a VE kernel (function) inside a *VeoLibrary*. Functions can be located inside a library by invoking explicitly the *VeoLibrary* object's *find_function()* method, or implicitly, by accessing an attribute of the *VeoLibrary* that is named like the expected function.

Memory buffers can be allocated inside the *VeoProc*, they are represented by *VEMemPtr* objects. These objects can be passed as arguments to synchronous or asynchronous VE-VH memory copying calls. Symbols inside *VeoLibrary* libraries loaded into the *VeoProc* can also be represented as *VEMemPtr* objects.

Inside the *VeoProc* process one or more *VeoCtxt* objects correspond to the actual worker threads on the VE. These objects are created through the *open_context()* method of *VeoProc* and wrap the opaque context structure described in Sect. 4.1.

A *VeoFunction* invoked on a *VeoCtxt* returns a *VeoRequest* object. The *Veo-Request* can be waited for (synchronous wait) with the method *wait_result()* or "peeked" at for asynchronous checks with *peek_result()*. The asynchronous VE function invocation is more seamless than in VEO's C-API, without the usage of the arguments object. The first argument of a function is the context on which it shall be executed, followed by the function's arguments. Due to the ambiguity of the Python data types, the argument and result types must be declared by the function's *args_type* and *ret_type* methods.

With *OnStack* objects it is possible to pass in and out arguments that need to be accessed by reference. Python objects that support the buffer interface are supported as arguments of *OnStack*. The initialization syntax is:

```
OnStack(buff, [size=...], [inout=...])
```

with the arguments:

- **buff**: is a python object that supports the buffer interface and is contiguous in memory.
- **size**: can limit the size of the transfer. If not specified, the size of the buffer is used.
- **inout**: the scope of the transfer, can be VEO_INTENT_IN, VEO_INTENT_OUT or VEO_INTENT_INOUT.

5.2 Building VE Offloadable Kernels

Simple VE kernels can be represented as strings inside Python and use the *VeBuild* class for compiling and building the dynamically shared VE library or veorun helper executable. This is an easy way to start experimenting with code on the Vector Engine.

After instantiating a *VeBuild* object one can define an arbitrary number of VE kernel source fragments by calling the methods *set_c_src()*, *set_cpp_src()* or *set_ftn_src()* for, respectively, C, C++ and Fortran sources. Each method takes the arguments

- **label**—a string with a unique label for the code fragment, e.g. its function name,
- **content**—a string containing the source code,
- **flags**—optional argument with compiler flags specific to the current code fragment,
- **compiler**—optional argument specifying the path to the compiler for the current code fragment.

In order to keep the current directory clean, one can specify a build directory by calling the *set_build_dir()* method. Finally, building the dynamically shared object is being done by invoking the *build_so()* method, which again allows for specifying compiler flags, libs, or a linker. Building a statically linked veorun helper is done by *build_veorun()*. Both build methods return the path to the built library or executable, which can be passed to the *VeoProc* methods.

5.3 PyVEO Example

The following example is a trivial piece of code using many of the PyVEO features.
A VE kernel that computes the average of an array of doubles is defined in a
VeBuild object. A shared object is created and its path is returned in the variable
ve_so_name.

An offloading process and context are created in proc and ctxt, then the VE
shared object is loaded and represented by the lib object. The *average* VE kernel's
argument and result types are set in the following two lines.

A random numpy array is created in the variable a, it's average is computed and
printed on the VH with the numpy average method, then an asynchronous request
req is created and submitted to the VE. The content of the array a is transfered
over the stack and passed into the VE kernel by reference through *OnStack*.

Finally the result avg is retrieved by waiting synchronously for the request to
finish.

```
1   import os
2   from veo import *
3
4   bld = VeBuild()
5   bld.set_build_dir("_ve_build")
6   bld.set_c_src("_average", r"""
7   double average(double *a, int n)
8   {
9       int i;
10      double sum = 0;
11
12      for (i = 0; i < n; i++)
13          sum += a[i];
14
15      return sum / (double)n;
16  }
17  """)
18  ve_so_name = bld.build_so()
19
20  # VE node to run on, take 0 as default
21  nodeid = os.environ.get("VE_NODE_NUMBER", 0)
22
23  proc = VeoProc(nodeid)
24  ctxt = proc.open_context()
25  lib = proc.load_library(os.getcwd()+"/"+ve_so_name)
26  lib.average.args_type("double *", "int")
27  lib.average.ret_type("double")
28
29  n = 100000 # length of random vector: 100k elements
```

```
30  a = np.random.rand(n)
31  print("VH numpy average = %r" % np.average(a))
32
33  # submit VE function request
34  req = lib.average(ctxt, OnStack(a), n)
35
36  # wait for the request to finish
37  avg = req.wait_result()
38  print("VE kernel average = %r" % avg)
39  del proc
```

6 Conclusion and Outlook

With VEO the SX-Aurora Tsubasa has received the basic mechanisms to expand the pool of usage models to a pure accelerator model with asynchronous VE kernel offloading while the VH is running the main program's threads. This report has introduced the implementation and API of VEO, which is in some extend similar to OpenCL and CUDA, but differs for example in the detail that offloaded VE kernels can call almost any systemcall. This detail makes the porting and hybridization of programs much easier because it allows for rather feature-rich offloaded code fragments.

VEO has been picked up by a set of projects which are either using it or implement different offloading paradigms with its help.

The TENSORFLOW port for VE [9] is offloading the neural networks computations to the VE while keeping the frontend of TENSORFLOW on the VH. This made porting much easier than a full native VE port.

Another AI application is SOL [10] which optimizes neural networks for PyTorch, TENSORFLOW and MXNET, compiles the optimized nets just-in-time for the VE and uses VEO for offloading the heavy neural network work while keeping unmodified plain x86_64 frontends on the VH. Right now SOL enables the VE to use three major AI frameworks without actually modifying them.

The Heterogeneous Active Messages (HAM) project [11] implements a C++ template library that uses VEO and VH-SHM mechanisms on the vector engine for implementing a low latency and high bandwidth framework for hybrid programming.

Work is being done to support the OpenMP target directive to offload to VE targets [12] by using VEO, which makes hybrid programming for C only require some additional *#pragma* directives.

Python bindings for VEO are available with the PyVEO project [8] and enable users to call VE kernels from inside Python programs.

This first implementation of VEO brought novel functionality to the SX-Aurora Tsubasa and helped porting various applications. At the same time it uncovered limitations in VEOS as well as in the implementation design, itself. Just recently, with VEOS 2.1.3, we were able to remove the limitation that proc instances could only be created from the main VH thread, not from child threads. The VEOS development is on the way to enable the simultaneous use of multiple VEs from one VEO process. In a project we work at improving the VH-VE transfer performance by switching to user DMA descriptors. Finally, the latency penalty due to the VEO request queue being placed on the VH side will need to be tackled soon, the HAM project showed a possible and promising approach.

References

1. Yamada, Y., Momose, S.: Vector engine processor of NEC's brand-new supercomputer SX-Aurora TSUBASA. In: Proceedings of A Symposium on High Performance Chips, Hot Chips (2018). https://www.hotchips.org [last visited: 05/19]
2. Komatsu, K., Momose, S., Isobe, Y., Watanabe, O., Musa, A., Yokokawa, M., Aoyama, T., Sato, M., Kobayashi, H.: Performance evaluation of a vector supercomputer SX-Aurora TSUBASA. In: Proceedings of the International Conference for High Performance Computing, Networking, Storage, and Analysis (SC '18), Article 54, 12 pp. IEEE, Piscataway. https://doi.org/10.1109/SC.2018.00057
3. Stone, J.E., Gohara, D., Shi, G.: OpenCL: a parallel programming standard for heterogeneous computing systems. IEEE Comput. Sci. Eng. **12**(3), 66–73 (2010)
4. Nickolls, J., Buck, I., Garland, M., Skadron, K.: Scalable parallel programming with CUDA. ACM Queue **6**(2), 40–53 (2008)
5. OpenMP Architecture Review Board: OpenMP Application Program Interface. Version 4.0, July 2013, https://www.openmp.org/wp-content/uploads/OpenMP4.0.0.pdf [last visited: 07/19]
6. The OpenACC Application Program Interface, Version 1.0, November 2011. https://www.openacc.org/sites/default/files/inline-files/OpenACC_1_0_specification.pdf [last visited: 07/19]
7. Vector Engine Offloading github repository. https://github.com/veos-sxarr-NEC/veoffload
8. PyVEO github repository. https://github.com/SX-Aurora/py-veo
9. TENSORFLOW-VE github repository. https://github.com/sx-aurora-dev/tensorflow
10. Weber, N.: Sol: Transparent neural network acceleration platform. In: Proceedings of Super-Computing (SC '18) (2018). https://sc18.supercomputing.org/proceedings/tech_poster/poster_files/post142s2-file3.pdf
11. Noack, M., Focht, E., Steinke, T.: Heterogeneous active messages for offloading on the NEC SX-Aurora TSUBASA. In: Proceedings of the 2019 IEEE International Parallel and Distributed Processing Symposium Workshops (IPDPSW), pp. 26–35 (2019). https://doi.org/10.1109/IPDPSW.2019.00014
12. Aurora OpenMP Offloading Documentation (2019). https://rwth-hpc.github.io/sx-aurora-offloading

Potential of LLVM for SX-Aurora

Simon Moll, Matthias Kurtenacker, and Sebastian Hack

Abstract The NEC SX-Aurora TSUBASA is a high-performance vector CPU for sustained simulation performance. The existing compiler toolchain for the SX-Aurora is comprehensive but also proprietary restricting its use in research and confining its development to internal teams at NEC. In recent years, the open source LLVM compiler infrastructure has seen significant support and contributions by major players such as NVIDIA, AMD, ARM, Intel, Apple and Google. These employ LLVM in their official toolchains, GPU driver stacks and mission-critical infrastructure. Likewise, many compiler research labs have adopted LLVM for its accessibility, robustness and permissive license. Recently, the LLVM community has been discussing an extension for scalable vector architectures (LLVM-SVE), which feature an active vector length just as the SX-Aurora does. In this paper, we will discuss the potential of LLVM for the NEC SX-Aurora. The Compiler Design Lab at Saarland University is working with NEC on an LLVM-SVE backend for the SX-Aurora.

1 Introduction

The NEC SX-Aurora TSUBASA (SX-AT) stands in a long line of Vector CPUs since the first Cray vector processors. Vector processing is getting more traction beyond SX-AT. It is the processing model of recent ARM ISAs (Helium, SVE, SVE2) and the RISC-V V extension.

This renewed popularity is driven mostly by two factors: the energy efficiency of the vector processing paradigm and the abundance of parallelism in modern HPC and Machine Learning codes.

The programming environment and tooling for SX-Aurora has been proprietary, hindering innovation by parties outside of NEC. This has recently changed with

S. Moll (✉) · M. Kurtenacker · S. Hack
Saarland University, Saarbrücken, Germany
e-mail: simon.moll@emea.nec.com; kurtenacker@cs.uni-saarland.de; hack@cs.uni-saarland.de

© Springer Nature Switzerland AG 2020
M. M. Resch et al. (eds.), *Sustained Simulation Performance 2018 and 2019*,
https://doi.org/10.1007/978-3-030-39181-2_10

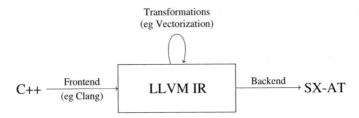

Fig. 1 Basic compilation pipeline of LLVM

a new interest by NEC to open up the SX-Aurora ecosystem for open source development. To this end, an open-source compiler backend for SX-Aurora is currently under development [5] NEC is co-operating with the Compiler Design Lab at Saarland University to build an LLVM backend for the NEC SX-Aurora Vector CPU.

The backend is based on the production-quality and open-source LLVM compiler infrastructure.

In this paper, we will discuss the opportunities and challenges of adopting the LLVM compiler for SX-AT development.

2 The LLVM Compiler Infrastructure

LLVM [7] is one of the leading open source compiler frameworks in terms of adoption by industry and research. LLVM has a library-based design and is intended to mesh well with applications that want to leverage compiler technology [6].

The main program representation of LLVM is the LLVM Intermediate Representation (LLVM IR). Static compiler pipelines based on LLVM follow a simple structure as shown in Fig. 1. Compiler frontends translate high-level program code into IR, a pipeline of transformations to optimize the program in IR and finally backends translate IR into target machine code. LLVM has been successfully employed for targeting a diverse set of architectures, ranging from x86 and ARM CPUs to FPGAs and GPU architectures (NVIDIA, AMD). As of writing, the official LLVM repository alone ships backends for 18 different architectures.[1]

C,C++ Support
LLVM supports C and C++ through Clang, the official LLVM frontend for C-like languages in the LLVM project. The Clang frontend currently supports C++17 [3]. Clang is open source just like LLVM and also follows a modular design.

---

[1] https://github.com/llvm/llvm-project/tree/master/llvm/lib/Target.

Fortran Support

Fortran is currently supported through the Flang project, which was open sourced after initial closed development by NVIDIA/PGI. The Flang project describes its current status as "production quality" with support for "Fortran 2003, some features from Fortran 2008, and OpenMP." [4]. Due to its origins in a proprietary code base and development, the current Flang compiler does not offer the same modularity and clean design as the Clang frontend.

Motivated by the short comings of current Flang, F18, a clean slate re-design was proposed. F18 was chosen to become the official Fortran compiler of LLVM. Different to current Flang, F18 is an open source project from the start and coordinates closely with the LLVM community. The expectation is that F18 will replace current Flang by 2020, by which time F18 will also be re-named to Flang.

3 Challenges in Bringing LLVM to SX-Aurora

The development of LLVM was driven by support for processors with short SIMD instructions, such as ARM NEON, IBM AltiVec and x86 SSE/AVX. This has lead to limitations when targeting Vector CPUs with LLVM at its current state.

LLVM IR has a builtin fixed-length vector type that supports SIMD operations as shown in Fig. 2. For example, `<8 x double>` is the LLVM IR type for a `double` vector with eight elements. All arithmetic instructions operate on scalar types as well as on vector types. For example `%x = fadd double %x, %y` is a scalar addition and `%a = fadd <8 x double> %b, %c` is the element-wise addition of two 8 element vectors.

However, there are three major limitations in this design when targeting SX-AT.

1. Native LLVM vector instructions do not support predication. Basically all vector instructions on SX-AT support a mask argument. This is also true for more recent SIMD ISAs such as x86 AVX512. Vector masking is a hard requirement to support side-effecting operations such as memory accesses or trapping arithmetic (division).

2. Native LLVM vector instructions do not feature an Active Vector Length (AVL). As for predication, SX-AT supports an Active Vector Length on almost every vector instruction. This also holds for the RISC-V V extension [15]. As for vector masking, proper modeling of the AVL in LLVM is critical for side-effecting operations.

```
1       %x = fadd double %y, %z
```
(a)

```
1       %x = fadd <8 x double> %y, %z
```
(b)

Fig. 2 Native SIMD types and operations in LLVM. (a) Scalar IR. (b) SIMD IR

3. There is a read-only register in SX-AT that provides vector length of hardware registers, the Maximum Vector Length (MVL). Future versions of SX-AT could double the vector length and the value of MVL would double as a result. The SX-AT ISA allows for vector programs that exclusively rely on MVL to determine the hardware vector length at runtime. Programs of that kind are called vector-length agnostic since they do not require any fixed vector length (e.g. 256 for current-generation SX-AT).

SX-AT shares this property with the ARM Scalable Vector Extension (SVE) and also the RISC-V V extension [14]. ARM proposed LLVM-SVE to enable vector-length agnostic programming in LLVM. This could also be used for SX-AT, if vector-length agnostic programming is desired. Given that for current SX-AT $MVL = 256$, LLVM-SVE support is not strictly necessary to fully support SX-AT in LLVM.

4 LLVM-VP: Vector Predication for LLVM

Current LLVM IR does not natively support vector instructions with neither masking nor an active vector length.

There is a simple approach that promises to offer a short-cut to Active Vector Length support in LLVM. The idea is to add a compiler builtin, and intrinsic, to LLVM IR that sets the current Active Vector Length. Any vector instructions after the call should then operate under the AVL that was last set before it. We show an example of this in Fig. 3a.

```
1    call @set_avl(256)
2    %c = fadd <256 x double> %a, %b
3    call @set_avl(128)
4    %e = fdiv <256 x double> %d, %c
```

(a)

```
1    call @set_avl(256)
2    %c = fadd <256 x double> %a, %b
3    %e = fdiv <256 x double> %d, %c
4    call @set_avl(128)
```

(b)

```
1    %c = fadd <256 x double> %a, %b
2    %e = fdiv <256 x double> %d, %c
3    call @set_avl(256)
4    call @set_avl(128)
```

(c)

Fig. 3 Arithmetic instructions in LLVM IR can move freely across functions call. (**a**) AVL setting before arithmetic as intended by the programmer or IR generator. (**b**) A legal re-ordering. The division is additionally performed on elements 128–255 with implications to floating-point exceptions (division by zero) and performance (latency of vector instructions depend on AVL). (**c**) Another legal re-ordering. The values of %c and %e entirely depend on the (unknown) initial state of AVL

```
1    %c = llvm.vp.fadd.v256f64(%a, %b, %m, i32 256)
2    %e = llvm.vp.fdiv.v256f64(%y, %z, %m, i32 128)
```

Fig. 4 LLVM-VP version of snippet Fig. 3a. The AVL and masks are passed as arguments. Legal re-orderings can not affect the AVL or mask argument

There is a problem with that approach. Arithmetic instructions in LLVM IR do not depend on the system state in any way, they are pure. Hence, transformations in LLVM may freely move arithmetic instructions around any functions calls, including calls set_avl. This results in IR programs that are legal re-orderings but that are invalid from the point of intended semantics as shown in Fig. 3b, c.

The Vector Predication extension for LLVM (LLVM-VP) [8] is a proposal to add support for both. LLVM-VP introduces a set of compiler intrinsics (builtin functions) to LLVM IR. There is one intrinsic per vector instruction. Different to the vector instructions, every LLVM-VP intrinsic has an explicit value argument for the Active Vector Length and the vector mask. The implicit dependence on the AVL register turns into an explicit value dependence (Fig. 4).

5   The Region Vectorizer

The Region Vectorizer [9] (RV) is a vectorization plugin for LLVM IR. RV is available on github.[2] RV is an enabling technology that can be used to build vectorizing compilers based on LLVM. For example, RV can be used to implement a vectorizing OpenCL driver similar to the Intel OpenCL Implicit Vectorization Module [10, 13].

RV vectorizes outer-loops and whole functions. In case of an outer-loop, it performs outer-loop vectorization. For the latter, it vectorizes entire scalar functions to process vector inputs. RV supports arbitrary reducible control-flow within those loops or functions. By that virtue, RV enables a programming model in C,C++ code that is ISPC-like [12] and similar to programming CUDA kernels [11].

In contrast, the vectorizers that are part of LLVM (LoopVectorizer and Unroll-and-jam) do not support any control-flow inside the vectorized region.

We provide a case study to demonstrate the implications of standard RV for SX-AT. Beyond this paper, there already exists published work on multi-dimensional vectorization for SX-AT using *TensorRV*, a version of the Region Vectorizer.

---

[2]Region Vectorizer. https://github.com/cdl-saarland/rv.

Fig. 5 Wavefront intrinsics

```
1  int foo(double v) {
2    double v = A[i];
3    if (any(v < 0.0)) {
4      // branch taken if for any thread v < 0.0
5      return -1;
6    } else {
7      return 1;
8    }
9  }
10
11 int foo_v2(double2 v) { /* generated by RV */ }
12 // foo_v2(<-1.0, 0.0>) = -1
13 // foo_v2(<1.0, 2.0>) = 1
```

5.1   Case Study: Tree Traversal Codes

The Region Vectorizer vectorizes arbitrary reducible control-flow and supports wavefront intrinsics (horizontal operations). In this case study, we will demonstrate the performance implications of these two capabilities.

Wavefront intrinsics [2] are functions that are compute one result for all threads that enter them. Consider the example in Fig. 5. The any function is an intrinsic that evaluates to true for all threads if its argument is true holds for any thread. In this example, the function foo is vectorized for two threads into a new function foo_v2. The scalar argument double becomes double2. Each lane holds the value of v for the respective thread. The examples below the vector function signature foo_v2 show how the vectorized version behaves.

Figure 6 shows a simple binary tree search for elements of the array Q in the tree given by node. In the classic thinking of loop vectorization, we would write a loop around the binary tree search kernel and hope for the compiler to vectorize it. We can force the compiler to vectorize the loop with the use of pragmas (e.g. OpenMP #pragma omp simd)>.

As we can see, the vectorized code will contain slow vector gather accesses. This is because the threads traverse the tree depending on the element they are querying. Therefore the access behavior on the tree structure is highly data-dependent.

We will now see how wavefront intrinsics and vectorizing with control-flow can be used to create an optimized binary tree search. Figure 7 shows the code. There are two key differences. First, the new algorithm maintains a stack, which might be surprising given that this is not necessary for the regular binary tree search. The stack is used as a worklist to keep track of all the nodes that still need to be visited.

Second, the algorithm uses the any-wavefront intrinsic. If any thread needs to visit a left or right child of the current node, that node is put on the stack for all threads to visit. The advantage lies in the fact that the binary tree nodes can be handled entirely in scalar registers. In the RV-generated version, vector code is only used in connection with the different query elements (elem).

Figure 7 is an example of a speculative traversal code [1].

```
1  // Binary tree node
2  struct Node {
3    double data;
4    int left;
5    int right;
6  };
7
8  // [..]
9
10 // Binary tree search
11 #pragma omp simd
12 for (int tid = 0; tid < 256; ++tid) {
13   int next = 0;                      // vector value
14   const double elem = Q[tid]; // vector load
15
16   while (next >= 0) {
17     double label = nodes[next].data; // vector gather
18     if (elem < label) {
19       next = nodes[next].left; // vector gather
20     } else if (label < elem) {
21       next = nodes[next].right; // vector gather
22     } else {
23       Result[tid] = next; // vector store
24       continue;
25     }
26   }
27   Result[tid] = -1; // vector store
28 }
```

Fig. 6 Binary tree search for 256 elements. Annotated is the kind of memory accesses that will be used when the loop is vectorized

The `while` loop in Fig. 6 will take as many iterations as the longest path taken by any thread. Figure 7 version will execute the `while` loop once for every node that will be visited.

The traversal of the binary tree is handled by the Scalar Processing Unit (SPU), which has an L1 cache one core. In contrast, the traversal of the data structure is handled entirely by the Vector Processing Unit (VPU) in Fig. 6. It uses less efficient memory accesses and the VPU has no cache close to core.

If the threads visit almost the same nodes in the traversal, Fig. 7 can outperform the standard loop version because of its more efficient traversal scheme. If the threads diverge too early in their traversal, the speculative version becomes inefficient because it does not benefit from the node stack anymore. In the extreme case, the speculative version degrades to scalar performance. Then, every iteration of the `while` will only be relevant for a single lane since only one thread actually needs to visit that node.

5.1.1 Results

We evaluate the performance of tree traversal codes [9] with control-flow and wavefront intrinsics on SX-AT. For these experiments, we use the LLVM-VP

```
1  // SPMD binary tree search with speculative traversal.
2  void
3  search(Node * nodes, double * Q, int * Result, int tid) {
4    // traversal stack
5    int stack[512];                    // shared across threads
6    stack[0] = 0; int top = 1;
7
8    const double elem = Q[tid]; // vector load
9    int result = -1;
10
11   while (top > 0) {
12     int next = stack[--top]; // scalar load
13     double label = nodes[next].data; // scalar load
14     int right = nodes[next].right; // scalar load
15     int left = nodes[next].left; // scalar load
16
17     if (label == elem) {
18       result = next;
19       break;
20     }
21     if (any(elem < label) && left > 0)
22       stack[top++] = left; // scalar store
23     if (any(label < elem) && right > 0)
24       stack[top++] = right; // scalar store
25   }
26   Result[tid] = result; // scalar store
27 }
28
29 // Vectorized binary tree element search for 256 elements.
30 void
31 search_v256(Node * nodes, double * Q, int * Result, int tid) {
32   // Generated by RV from LLVM IR of "search" function.
33 }
```

Fig. 7 The Region Vectorizer enables ISPC-like programming in any language that translates to LLVM IR

versions of the Region Vectorizer [9] and the experimental SX-AT backend.[3] The traversal codes operate on `double`-type data.

Platforms

All results are for a single core on either the Vector Host (VH) or the SX-AT Vector Engine (VE).

- **VE** A single NEC Aurora TSUBASA Vector Engine 10B model, with 1.4 GHz clock frequency.
- **VH** Intel(R) Xeon(R) Gold 6126 CPU (Vector Host).

---

[3]https://github.com/cdl-saarland/llvm-aurora-dev/tree/develop_cdl.

Configurations and Compilers

We tested the following configurations for each benchmark:

- **VH-GCC** GCC 7.3.1, relying on automatic loop vectorization (no wavefront intrinsics or speculative traversal).
- **VH-Clang** Clang (based on Clang/LLVM repositories as of March, 2019). Using LLVMs own vectorizers.
- **VH-Clang + RV** LLVM + Clang + RV. (Wavefront intrinsics and speculative traversal).
- **VE-NCC** NCC 2.1.1, relying on automatic loop vectorization (no wavefront intrinsics or speculative traversal).
- **VE-Clang** LLVM-VE backend + Clang. Scalar traversal codes (LLVM can not vectorize the outer-loops by its own means, no wavefront intrinsics, no speculative traversal).
- **VE-Clang + RV** LLVM-VE backend + LLVM-VP + Clang + RV. Tree traversal codes vectorized by RV (Wavefront intrinsics and speculative traversal).

Benchmarks

The traversal codes and inputs comprised the following:

- **bintree** Binary tree search as in Fig. 7 for the wavefront version and Fig. 6 for the loop version. 2^18 tree elements.
- **kmeans-kd** Kmeans algorithm for 128 clusters on 10^6 many random 2D coordinates. The kd-Tree is built for the 128 cluster centers to accelerate the 1-NN search.
- **nn-kd** 1-NN search over 10^6 many random 2D coordinates using a kd-tree.
- **pc-kd** Point Correlation search over 10^6 random 2D coordinates (kd-Tree).
- **nn-vp** 1-NN search over $10^6 A$ random 2D coordinates (Vantage Point tree).
- **xsbench** Inner loop of the DoE SXBench proxy application (nuclide grid option). Binary tree search on an array of 10^7 elements.

Figure 8 shows the results normalized to speedups over the VH-GCC configuration. We make the following observations:

Speculative traversal can enable significant speedups on both platforms, VH and VE. We see speedups of up to $\times 10.1$ for kmeans-kd on the VH and up to $\times 1.78$ on the VE.

However, the speculation does not always pay off. The speculative RV-versions of NN-VP perform worse than the scalar version on the VH. Also, the scalar VE-Clang code is less efficient than the code that NCC generates.

GCC does not seem to vectorize the traversal loops. We know that the VH-Clang version does not vectorize the traversal loops and performance is close to GCC.

The difference between scalar code and speculative vector code is more pronounced on the VE than on the VH. This is unsurprising given the smaller vector

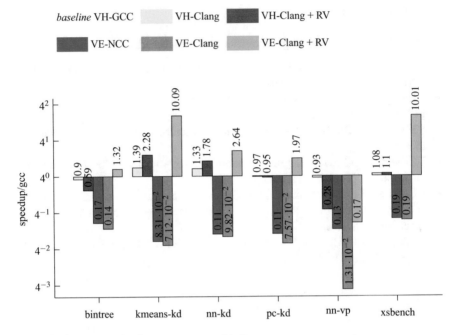

Fig. 8 Performance evaluation on tree traversal codes

size on the VH (16 `double` elements) whereas the VE has the potential to process 256 queries at once.

The results show that the VE-Clang+RV programming model is able to outperform the classic NCC outer-loop vectorization model. RV operates on speculative traversal codes, whereas NCC is given the loop version, which puts NCC at a disadvantage in this benchmark. However, it should be noted that NCC does not support the wavefront programming model the way RV enables it for LLVM.

6 Conclusion

LLVM is a state of the art compiler framework for static compilation. In this paper, we discuss the potential as well as the challenges that need to be addressed to fully leverage LLVM for SX-AT development.

The LLVM ecosystem benefits from its modular structure. It feature the state-of-the-art Clang frontend for C and C++ as well as serious efforts to bring Fortran support to LLVM.

At the same time, the SIMD support in current LLVM is insufficient to fully target the vector ISA of SX-AT. Efforts such as the LLVM Vector Predication extension show that these concerns are being addressed. The existence of proposals like LLVM-VP shed light on the benefits of an open source compiler infrastructure.

We show in a study of tree traversal codes that the Region Vectorizer for LLVM enables a new programming model for SX-AT. The speculative programming model of RV unlocks performance on SX-AT that is inaccessible from the proprietary NEC compiler toolchain.

References

1. Aila, T., Laine, S.: Understanding the efficiency of ray traversal on GPUs. In: Proceedings of the Conference on High Performance Graphics 2009, HPG '09, pp. 145–149. ACM, New York (2009)
2. AMD. Graphics Core Next Architecture, Generation 3. http://developer.amd.com/wordpress/media/2013/12/AMD_GCN3_Instruction_Set_Architecture_rev1.1.pdf (2016). Accessed 26 Jun 2019
3. C++ Support in Clang. https://clang.llvm.org/cxx_status.html. Accessed 26 Jun 2019
4. Flang and F18. https://github.com/flang-compiler/flang/wiki. Accessed 26 Jun 2019
5. Ishizaka, K., Marukawa, K., Focht, E., Moll, S., Kurtenacker, M., Hack, S.: NEC SX-Aurora - A Scalable Vector Architecture (2018). https://compilers.cs.uni-saarland.de/papers/nec_poster_llvmdev18.pdf
6. Lattner, C.: Llvm. http://www.aosabook.org/en/llvm.html (2019). Accessed 10 July 2019
7. Lattner, C., Adve, V.S.: LLVM: A compilation framework for lifelong program analysis & transformation. In: 2nd IEEE/ACM International Symposium on Code Generation and Optimization (CGO 2004), San Jose, pp. 75–88 (2004)
8. Moll, S.: (D57504) RFC: Prototype & Roadmap for Vector Predication in LLVM. https://reviews.llvm.org/D57504. Accessed 26 Jun 2019
9. Moll, S., Hack, S.: Partial control-flow linearization. In: Proceedings of the 39th ACM SIGPLAN Conference on Programming Language Design and Implementation, PLDI 2018, Philadelphia, pp. 543–556 (2018)
10. Munshi, A.: The OpenCL Specification. In: 2009 IEEE Hot Chips 21 Symposium (HCS), pp. 1–314. IEEE, Piscataway (2009)
11. Nickolls, J., Buck, I., Garland, M., Skadron, K.: Scalable parallel programming with CUDA. ACM Queue **6**(2), 40–53 (2008)
12. Pharr, M., Mark, W.R.: ispc: A SPMD compiler for high-performance CPU programming. In: 2012 Innovative Parallel Computing (InPar), pp. 1–13 (2012)
13. Rotem, N.: Intel Opencl Implicit Vectorization Module (2011)
14. Stephens, N., Biles, S., Boettcher, M., Eapen, J., Eyole, M., Gabrielli, G., Horsnell, M., Magklis, G., Martinez, A., Pr'emillieu, N., Reid, A., Rico, A., Walker, P.: The ARM scalable vector extension. IEEE Micro **37**(2), 26–39 (2017)
15. Working draft of the proposed RISC-V V vector extension. https://github.com/riscv/riscv-v-spec (2019). Accessed 26 Jun2019

Bad Nodes Considered Harmful: How to Find and Fix the Problem

Marco Seiz, Johannes Hötzer, Henrik Hierl, Stefan Andersson, and Britta Nestler

Abstract Large, distributed systems of computing units are the current state of the art for conducting high-performance computing. With large systems comes an increasing chance of failure of any component in the system, necessitating research as how to cope with failure. Failures may manifest as compute nodes shutting down, but also in differing performance among compute nodes. This chapter concerns itself with investigating a recent occurrence of the latter and how to avoid this in large scale runs.

1 Overview of the Problem

High-performance computing is used to accelerate the development of various applications ranging from weather predictions over medical imaging to materials simulations. In order to calculate large domains in reasonable times, many (dis-

The authors "Marco Seiz and Johannes Hötzer" contributed equally.

M. Seiz
Institute of Applied Materials (IAM), Karlsruhe Institute of Technology (KIT), Karlsruhe, Germany

J. Hötzer (✉) · B. Nestler
Institute of Applied Materials (IAM), Karlsruhe Institute of Technology (KIT), Karlsruhe, Germany

Institute for Digital Materials (IDM), Hochschule Karlsruhe — Technik und Wirtschaft (HSKA), Karlsruhe, Germany
e-mail: johannes.hoetzer@kit.edu

H. Hierl
Institute for Digital Materials (IDM), Hochschule Karlsruhe — Technik und Wirtschaft (HSKA), Karlsruhe, Germany

S. Andersson
Amazon Web Services (AWS)
e-mail: lrande@amazon.com

© Springer Nature Switzerland AG 2020
M. M. Resch et al. (eds.), *Sustained Simulation Performance 2018 and 2019*,
https://doi.org/10.1007/978-3-030-39181-2_11

tributed) compute units are necessary. Each compute unit however has its own individual error probability p, with the error manifesting as e.g. complete shutdown or performance deficits. With N connected compute units, the probability of at least one error occurring is $1 - (1 - p)^N$, hence the chance of at least one error occurring sharply rises with the system size N. Schroeder and Gibson [1] investigated the node shutdown failures of different large systems over multiple years, ranging from 3 to 0.1 failures per year per processor. Consider the probability of at least one error occurring on any day reaching 50%, then we would require between 84 processors and 2530 processors for the systems investigated by Schroeder and Gibson. The HLRS Hazel Hen system considered in this chapter has a failure rate of less than 0.000 56 failures per year per processor. Applying the same consideration for the Hazel Hen, this would require 451 796 processors. The estimates based on Schroeder and Gibson's data are likely below the number of processors in the considered systems, implying daily failures, but for the Hazel Hen the number of processors is very roughly about three times the number of available processors, i.e. a failure would occur only every 3rd day.

When node shutdown occurs, parallel applications employing the message passing interface (MPI) will usually terminate the parallel application. To safeguard against this and continue the programs regular checkpoints can be stored to disk. Recently extensions to MPI have been discussed under the umbrella of resilience to recover the program execution [2–4].

Besides node shutdown, nodes may also perform below specifications. When such nodes are connected, their inhomogeneous performance can impact total performance as well as scalability. The performance deficit incurred by inhomogeneously performing compute units has been investigated previously by Acun et al.[5, 6]. Therein they showed that environmental conditions such as temperature or power draw induce clock frequency variations due to Turbo Boost. Mitigations for clock frequency variation suggested by Acun et al. include disabling Turbo Boost, frequency pinning and dynamic load balancing. In his PhD thesis, Acun [7] investigated the reasons for frequency variation more closely, concluding that both chip temperature and power draw are significant factors. Inadomi et al. [8] considered performance variation under the lens of power constraints and manufacturing variability. As not all chips are born equal, the chips get binned according to their performance characteristics after manufacturing. When a power draw constraint is imposed on chips of such a bin, performance is affected drastically [9]. Further investigations include [10, 11].

These kinds of performance variations can also be observed on the HLRS Hazel Hen system, as shown in the histogram of node performance in Fig. 1. While the variation seems to be limited with around ± 30 GFLOP/s ($\pm 4\%$) around the median in these measurements, the worst compute units may be much worse than this. Case in point being the weak scaling results depicted in Fig. 2, which is based on a 3D domain decomposition of the explicit time integration of two coupled partial differential equations. While the efficiency is almost ideal up to 100,000 cores, a sharp drop occurs beyond this. In the rest of this chapter, we will specify the

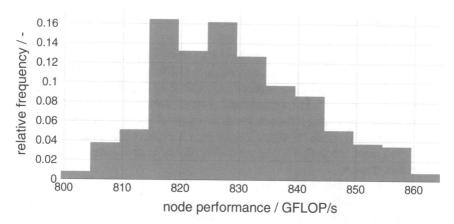

Fig. 1 Histogram of per-node HPL performance results on the Hazel Hen. The minimal performance is 801 GFLOP/s and the maximal performance is 861 GFLOP/s with the median performance at 828 GFLOP/s

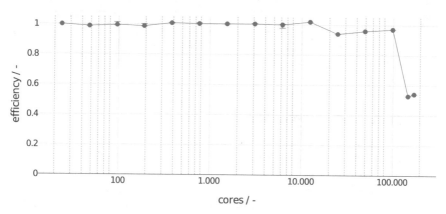

Fig. 2 Weak scaling behavior on the Hazel Hen for a forward time central space scheme of a system of two coupled partial differential equations. After almost ideal scaling up to 100,000 cores there is a sharp drop in efficiency. This chapter is concerned with the reason for this sudden drop

experimental conditions for this, analyze the root cause of this performance drop and show solutions for this kind of problem.

2 Environment

In this section we will specify the computing environment as well as the code used for the experiments.

2.1   The HLRS Cray XC40 Hazel Hen

For our measurements we used the Hazel Hen supercomputer [12] located at the High-Performance Computing Center Stuttgart (HLRS). The Cray XC40 system consists of 7712 nodes with 2 sockets each with a 12 core Intel Xeon CPU E5-2680 v3, each of which has a theoretical peak performance of 480 GFLOP/s per socket. For the interconnect four nodes share a Cray Aries network chip which are connected using a DragonFly network topology. Hazel Hen employs the Application Level Placement Scheduler to map parallel applications onto compute nodes with the aprun command.

For the Hazel Hen system, Cray uses several tools to monitor the system and pro-actively removes any suspicious node from the production pool before it causes problems. These tools either monitor the whole system, like the 'Lightweight Log Manager' which collects all log files from the nodes or a monitor, which constantly checks the number of ECC error of the DIMMS on all nodes. Other tools work on a job level, like the 'Node Health Checker' (NHC), which performs several health checks at the end of a job. The combination of all these efforts is that Hazel Hen, on average, experiences less than 2 nodes failures on running jobs per week. Assuming that the whole machine is in full use this translates to a failure rate of less than 0.00056 failures per year per processor as mentioned in the introduction.

2.2   The PACE3D Framework

The massive parallel framework PACE3D [13] is a multi-physics solver to study materials science processes with the phase-field method [14]. Parallelization of the code is achieved with the message passing interface (MPI). Each rank is assigned its own rectangular subdomain with ghost layers depending on the discretization.

For the measurements shown in this work, two coupled partial differential equations (ϕ, μ) are used based on the model in [15]. In Algorithm 1 the general structure of the calculation and communication is shown. The equations are implemented as separate sweeps using a src field of the current time step and dst field to store the new time step. Discretizing these with the classical forward-time-central-space scheme requires one cell thick ghost layers in each direction. Each equation update depends on the direct neighboring cells (D3C7) of its own src field as well as the center cell (D3C1) of the other field. Due to this kind of dependency, calculation and communication of both equations can be overlapped, which is realized with non-blocking sends.

Additionally to the parallelization with MPI and overlapping communication, optimizations on various levels are performed, with the entire application showing up to 32.5% of the theoretical single core peak performance. The code was compiled with GCC 8.2.0 with cray-mpich 7.7.4 providing the MPI implementation.

Algorithm 1 Time step with overlapping communication

1: END: COMMUNICATION GHOST LAYERS (ϕ_{src})
2:   $\phi_{dst} \leftarrow \phi$**-calculation** $\left(\phi_{src}, \mu_{src}\right)$ //Stencil dependencies: D3C7 (ϕ_{src}) and D3C1 (μ_{src})
3:   BOUNDARY CONDITIONS(ϕ_{dst})
4: START: COMMUNICATION GHOST LAYERS (ϕ_{dst})
5: END: COMMUNICATION GHOST LAYERS(μ_{src})
6:   $\mu_{dst} \leftarrow \mu$**-calculation** $\left(\mu_{src}, \phi_{src}, \phi_{dst}\right)$ //Stencil dependencies: D3C7 (μ_{src}), D3C1 (ϕ_{src})
     and D3C1 (ϕ_{dst})
7:   BOUNDARY CONDITIONS (μ_{dst})
8: START: COMMUNICATION GHOST LAYERS (μ_{dst})
9:   SWAP $\phi_{src} \leftrightarrow \phi_{dst}$ and $\mu_{src} \leftrightarrow \mu_{dst}$

Timers based on `clock_gettime()` are employed to measure runtimes. These are set around the ϕ computation, the μ computation, the waits on the non-blocking communication, the initialization, as well as around the entire application. The sum of the individual timers compares reasonably with the timer for the entire application. Per-process timing results are stored in an SQL database. During the scaling runs no data was written to disk. By appropriate construction of the simulation domain, the processes were ensured to be load balanced. A sufficient number of iterations was calculated to reach steady-state performance. The frequency of the processors was pinned to 2.5 GHz.

3 Analysis of the Problem

To analyze the performance drop of Fig. 2 we used the individual timer data stored in the SQL database. In Fig. 3 the total time for the ϕ calculation, the μ calculation, the ϕ communication and the μ communication are plotted over the ranks. Approximately in the middle of the plot a clear increase of the two calculation times can be seen. The magnification in Fig. 3b reveals that exactly 24 cores are affected by this.

By employing the environment variable `MPICH_RANK_REORDER_DISPLAY`, which shows the mapping between rank IDs and the physical compute nodes, the 24 cores were found to be located on a single bad node. In a joint analysis with the HLRS it was found that this node was running at the minimum frequency of 1.2 GHz. As this was not considered an error by the system software at the time, it was not reported.

Once the bad node was determined, it was excluded from following runs by employing the aprun option `--exclude-node-list` to exclude it and allocating an additional node to compensate for the potentially bad node.

(a)

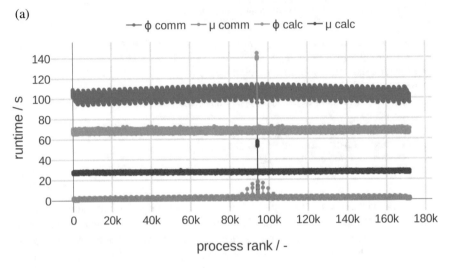

(b)

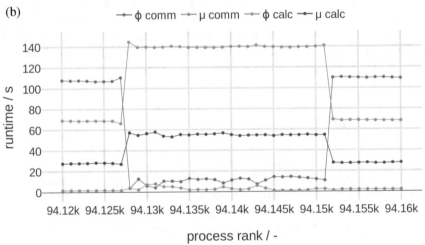

Fig. 3 Runtimes for the two kernels (ϕ calc, μ calc) and the two ghost layer exchanges (ϕ comm, μ comm) over the full machine and magnified around the peak. The timings show a clear increase of the calculation time for 24 cores which correspond to 1 node. This increase of calculation time causes all other communication time to increase dramatically. (**a**) Runtimes for all processes by application part. (**b**) Zoomed-in view of (**a**) around the abnormal processes

4 Solution

Following this manual solution, the bad node was excluded from the node pool at HLRS. In order to be resilient against further bad nodes, a script to detect these bad nodes was developed. A small benchmark (e.g. HPL) is calculated on each node, giving performance information per node (P_i) in terms of FLOP/s. The median of the performance (P_m) is determined and the relative difference in performance ΔP is given by:

$$\Delta P = (P_i - P_m)/P_m \tag{1}$$

If ΔP is negative and deviates by more than a threshold percentage from the median, the node is excluded. Algorithm 2 shows pseudocode for the principal design of a script to realize this.

Algorithm 2 Algorithm for filtering out bad nodes

$thresh \leftarrow -0.2$
for node in NODES **do**
 execute benchmark on each node, write to PERFFILE
end for
determine median performance P_m
for node, perf in PERFFILE **do**
 $\Delta P = (perf - P_m)/P_m$
 if $\Delta P < thresh$ **then**
  print node to file
 end if
end for
execute parallel application without bad nodes

Whenever this approach is applied, the job resources must contain more nodes than required for the application, as some nodes may be excluded by this method. When too many nodes are outside of this performance window, the parallel application may fail to start.

5 Conclusions

In this work we have shown how bad nodes can impact the performance as well as scalability of parallel codes. Based on per-process timers the bad node was identified and excluded in further runs. With this experience, a benchmark script was developed together with the HLRS to automatically exclude bad nodes. Finally, Cray has defined a test to check against nodes running at abnormally low frequency, allowing the HLRS to exclude these bad nodes without requiring user intervention. In the case that the computing center is not testing for performance errors, it is advisable to manually test large scale jobs for bad nodes.

Acknowledgements We are grateful to the computational resources on the Hazel Hen provided by the HLRS Stuttgart and their support. We thank Thomas Beisel and Bernd Krischok for the possibility to conduct multiple full machine runs. We also thank for the funding from BMBF at the project SKAMPY and for the funding received from the Deutsche Forschungsgemeinschaft under grant NE 822/9-2.

References

1. Schroeder, B., Gibson, G.: A large-scale study of failures in high-performance computing systems. IEEE trans. Depend. Secure Comput. **7**(4), 337–350 (2009)
2. Bland, W., Bouteiller, A., Herault, T., Bosilca, G., Dongarra, J.: Post-failure recovery of mpi communication capability: design and rationale. Inter. J. High Perform. Comput. Appl. **27**(3), 244–254 (2013)
3. Losada, N., Cores, J., Martín, M.J., González, P.: Resilient MPI applications using an application-level checkpointing framework and ULFM. J. Supercomput. **73**(1), 100–113 (2017)
4. Kohl, N., Hötzer, J., Schornbaum, F., Bauer, M., Godenschwager, C., Köstler, H., Nestler, B., Rüde, U.: A scalable and extensible checkpointing scheme for massively parallel simulations. Inter. J. High Perform. Comput. Appl. **33**, 571–589, (2018). https://doi.org/10.1177/1094342018767736
5. Acun, B., Miller, P., Kale, L.V.: Variation among processors under turbo boost in hpc systems. In: Proceedings of the 2016 International Conference on Supercomputing, ICS'16, pp. 6:1–6:12. ACM, New York (2016)
6. Acun, B., Kale, L.V.: Mitigating processor variation through dynamic load balancing. In: 2016 IEEE International Parallel and Distributed Processing Symposium Workshops (IPDPSW), pp. 1073–1076. IEEE, Piscataway (2016)
7. Acun, B.: Mitigating variability in HPC systems and applications for performance and power efficiency. PhD Thesis, University of Illinois at Urbana-Champaign (2017)
8. Inadomi, Y., Patki, T., Inoue, K., Aoyagi, M., Rountree, B., Schulz, M., Lowenthal, D., Wada, Y., Fukazawa, K., Ueda, M., et al.: Analyzing and mitigating the impact of manufacturing variability in power-constrained supercomputing. In: SC'15: Proceedings of the International Conference for High Performance Computing, Networking, Storage and Analysis, pp. 1–12. IEEE, Piscataway (2015)
9. Rountree, B., Ahn, D.H., De Supinski, B.R., Lowenthal, D.K., Schulz, M.: Beyond DVFS: A first look at performance under a hardware-enforced power bound. In: 2012 IEEE 26th International Parallel and Distributed Processing Symposium Workshops & PhD Forum, pp. 947–953. IEEE, Piscataway (2012)
10. Chunduri, S., Harms, K., Parker, S., Morozov, V., Oshin, S., Cherukuri, N., Kumaran, K.: Run-to-run variability on Xeon phi based cray XC systems. In: Proceedings of the International Conference for High Performance Computing, Networking, Storage and Analysis, SC '17, pp. 52:1–52:13. ACM, New York (2017)
11. Marathe, A., Zhang, Y., Blanks, G., Kumbhare, N., Abdulla, G., Rountree, B.: An empirical survey of performance and energy efficiency variation on intel processors. In: Proceedings of the 5th International Workshop on Energy Efficient Supercomputing, E2SC'17, pp. 9:1–9:8. ACM, New York (2017)
12. CRAY XC40 Hardware and Architecture. https://kb.hlrs.de/platforms/index.php/CRAY_XC40_Using_the_Batch_System. Accessed March 2019
13. Hötzer, J., Reiter, A., Hierl, H., Steinmetz, P., Selzer, M., Nestler, B.: The parallel multi-physics phase-field framework pace3D. J. Comput. Sci. **26**, 1–12 (2018)
14. Nestler, B., Garcke, H., Stinner, B.: Multicomponent alloy solidification: phase-field modeling and simulations. Phys. Rev. E **71**(4), 041609 (2005)
15. Hötzer, J., Seiz, M., Kellner, M., Rheinheimer, W., Nestler, B.: Phase-field simulation of solid state sintering. Acta Mater. **164**, 184–195 (2019)

vTorque: Introducing Virtualization Capabilities to Torque

Nico Struckmann

Abstract The flexibility and portability commonly known from Clouds provide many benefits to users, software developers, administrators and data-center owners. With emerging technologies addressing today's major bottleneck of virtualization technologies, the virtual I/O, incentives for adoption of virtualization in HPC infrastructures arise. The advantages are manyfold for virtualization in HPC. Users can be served with flexible customized environments. Software developers can package applications with best matching dependencies. Administrators can upgrade or change their HPC infrastructure, i.e. the operating system, without impact on the applications served. Data-center owners can serve conflicting user groups and increase overall resource utilization by consolidating workloads with opposite characteristics. vTorque is an non-intrusive approach to introduce virtualization capabilities to the PBS based batch-system resource manager Torque. For traditional HPC infrastructures it enables cloud-like features, e.g. flexibility and portability, while maintaining the ability to run jobs on bare metal. vTorque further integrates available optimizations for virtual I/O throughout the whole stack, from the hypervisor to the guest level, as optional components.

1 Introduction

HPC and Clouds are quite opposite compute environments serving different purposes, having disjunct pros and contras.

HPC is usually a static environment in terms of operating system, kernel version, compute resource provisioning, and site specific properties, such as paths for homes and intermediate storage. Often needs of different user groups are in conflict to each other, e.g. the best matching kernel version. Additionally, HPC applications often require adaptation for each HPC environment. Resource allocation is usually user-

N. Struckmann (✉)
High Performance Computing Center Stuttgart (HLRS), Stuttgart, Germany
e-mail: struckmann@hlrs.de

© Springer Nature Switzerland AG 2020
M. M. Resch et al. (eds.), *Sustained Simulation Performance 2018 and 2019*,
https://doi.org/10.1007/978-3-030-39181-2_12

exclusive and may not provide best overall utilization. Compared to clouds HPC environments provide a higher performance, especially for parallel applications and thus reduce the time to result.

Cloud on the other hand is highly flexible and has no restrictions other than compatibility of the CPU architecture (e.g. ARM, x86). Developers can prepackage their applications with best matching dependencies. Applications are portable and can run without any adaptation in any other cloud environment as long as the CPU architecture matches. Another aspect of Cloud is, resources are usually shared and so workloads with different characteristics such as I/O intensive vs. CPU intensive vs. memory intensive applications could be deployed onto the same resources. Such consolidations increase resource utilization. Fault resiliency is provided by the help of automatic migration of running instances to another host, as well as suspend and restart of applications and services. This flexibility and resiliency, however, has its price in the shape of virtualization overheads.

The motivation for vTorque is to combine the best of both worlds, introduce the benefits of Clouds into HPC, with the target to gain high flexibility, to increase overall resource utilization, reduce administration and maintenance efforts, provide ability to serve conflicting user groups, while preserving the highest possible performance. While vTorque focuses in first place on introducing virtualization capabilities into HPC batch system resource manager Torque [1], several I/O optimizations provided by optional external components are supported.

2   vTorque

vTorque introduces virtualization capabilities to PBS based HPC resource manager Torque. It is comprised of a collection of non-intrusive bash scripts and several templates for user level script wrappers and virtual guests files. Plugins for various virtualized I/O optimizations are supported. It is integrated with Torque by the help of various root level and user level hooks for different phases of the workload management, as well as templates for virtual guest definitions and system customization.

Further, vTorque introduces two new command line interface (cli) tools, one for virtual job submission (vsub) and the other one for virtual guest image management (vmgr). Due to its non-invasive nature, it works smoothly with all recent and current Torque versions out of the box. vTorque's vsub supports qsub's cli and in addition comes along with various additional command line options related to virtual execution environment. The vsub cli has overriding capabilities for default values provided by admins, such as the image to use or default resource assignments.

2.1 Features

vTorque is covering all basic needs to run jobs as used to with Torque. For future work see Sect. 5, it outlines additional sophisticated features and functionality. The list of features comprise:

- Deploys job scripts transparently to the user in customized virtual compute node environments.
- Virtual node allocations have same hostname as their bare-metal node, but prefixed with a 'v', suffixed with the local vm number, e.g. '−1'.
- Multiple VMs per node supported.
- Pinning of virtual CPUs, including support for automated pinning with numad.
- Contextualization and customization of guests during instantiation by the help of NoCloud metadata technology.
- Root and user level VM pro/epilogue script support (for standard Linux guests) as known from Torque for bare metal compute nodes.
- Abstraction layer for environment specific properties to be defined by administrators, such as network adapters or file-system mount points, enabling portability between HPC and Cloud compute environments.
- Administrators define min/max and default values for their users, i.e. amount of virtual CPUs.
- Bare metal nodes and virtual guests share the same (network) file-systems for e.g. */home* and *workspace*.
- Clusterwide installations (usual in */opt*) are mounted as */opt-hpc* in guests to enable applications to make use of proprietary libs and commercial tools, as far as binary compatibility is given.
- Supports for different guest operating systems

  - Debian based Linux
  - RedHat based Linux
  - Unikernel OSv

- Several optional components available, e.g. for full stack monitoring and I/O optimizations. Please refer to Sect. 2.6.
- Configurable logging provided by *Log4Bsh* with several log levels (*ERROR/ WARN/INFO/DEBUG/TRACE*), filterable by component to debug and optional live console output.
- Compatible with Torque setups where a dedicated scheduler, e.g. Moab, is deployed.

2.2 Workflow

vTorque's non-invasive integration with Torque uses various script files run by Torque during workload deployment. These steps are used to prepare the compute node environment prior to a job allocation, as well as to clean it up afterwards. There

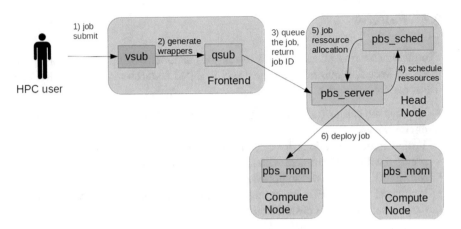

Fig. 1 Job submission with vTorque

are root level scripts provided for administrators, and also user level scripts hooks exist to execute logic before and after a batch job's execution.

Figure 1 illustrates Torque's runtime sequences for a batch job's deployment and execution, and how vsub comes into play for virtualized workloads.

1. User submits batch job script, but uses *vsub* instead of Torque's *qsub* cli.
2. Cli vsub generates first set of wrapper files for the user jobscript and user prologue. And subsequently passes the job submission request on to Torque, including all other received qsub arguments
   qsub -l prologue=<vmPrologue.sh> <other arg received> <jobscriptwrapper>
   At first *vsub* generates a random unique identifier (RUID) and wrapps the actual user job-script. This RUID is used to identify generated artifacts for each job during succeeding steps of the life-cycle. Torque's job ID is not yet assigned at that point in time.
3. Torque's qsub sends the wrapped job submission request to the *pbs_server* running on the head node.
4. Torque schedules the job.
5. And assigns the resource allocation.
6. Torque's compute node daemon *pbs_mom* deploys and executes the wrapped job.

The files generated by *vsub* cli are executed during different phases of Torque's deployment workflow. Since those wrappers are replacing wrapped Torque's original scripts, obviously they get executed also for bare-metal jobs. And add few seconds timeout overheads to each deployment, while waiting for VM job related files written by *vsub* on the frontend to appear on the shared file-system polled.

Figure 2 illustrates Torque's run-time sequences among allocated compute nodes *rank_0 - rank_n+1* for workload deployment and execution, and how those are wrapped by vTorque to enable virtualization capabilities.

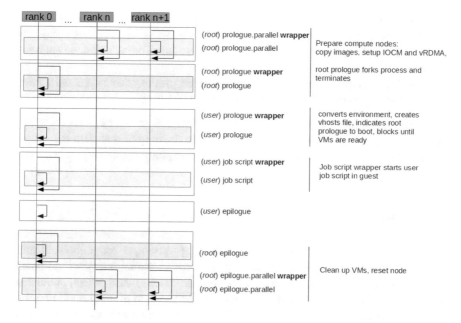

Fig. 2 vTorque workflow

1. Torque's root level *prologue.parallel* script is wrapped by vTorque's one and executed on all nodes, but the first (*rank_0*). It executes as first the original script to maintain all original functionality, followed by the initialization of optional modules. At the end it forks and returns the control flow, while waiting for the user level scripts to prepare contextualization of virtual guests.
2. The root level wrapper script for the first node (*rank_0*) is executed with same functionality as the *prologue.parallel* wrapper script.
3. User level wrapper script *vmPrologue.sh* is executed, which was submitted to Torque's *qsub* by vTorque's *vsub*. It copies VM images to and generates metadata files on the compute nodes, by the help of an additional script *vmPrologue.parallel.sh* executed on all nodes, including *rank_0*. A flag file is written indicating the forked root level prologue scripts, guests can now be instantiated. The *vmPrologue.sh* blocks until all VMs are reachable via SSH, creates the nodesfile for the guest environment and runs optional user prologues wrapped on baremetal and VM level.
4. Torque starts the wrapped job-script, which fetches Torque environment variables and adapts those as far as needed for the virtual guest. This comprises for example the *PBS_NODELIST* variable pointing to the nodes file with all hostnames allocated for the job. In a succeeding step, the job-script is executed via SSH on the first virtual node (representing *rank_0* in the virtual environment) and blocks until the job has finished.
5. Torque's optional user epilogue is run, no vTorque wrappers are involved.

6. Root level wrapper script *epilogue* is executed on the first node (*rank_0*) and resets the compute node. It stops the virtual guest(s), removes all optionally loaded components, removes temporary files and runs Torque's original *epilogue* script.
7. Root level wrapper script *epilogue* is executed on the other nodes with same functionality, except cleaning up on the shared file-system.

The virtual layer mirrors the concept of prologue and epilogues on root level, as it is available for the bare metal environment. By the help of metadata the scripts get executed during their instantiation and shutdown phase.

2.3   Artefacts

vTorque wraps Torque's scripts, such as the root prologue script, hooked into Torque's workflow to accomplish the management of virtual nodes within batch job life-cycles. vTorque's wrapper scripts contain various placeholders, populated at run-time during the workload deployment when required properties are present. Those populations depend on various PBS run-time environment variables, job script and vsub cli arguments, as well as administrator settings. As example the node list, amount of nodes, cpus per host, memory per host, guest image and optional components to load. In addition, there are XML domain template files [10] to define and customize virtual guest environments, based on qemu, libvirt and KVM. VM template files provide definitions of guest's devices, like file-system mount points and available devices to pass through. Combined with NoCloud metadata technology [9] for contextualization, for example to create the user's account and install security updates during VM instantiation, it is a flexible solution, utilizing state of the art Cloud technologies. vTorque offers a configuration file, enabling administrators to define default resources for job execution, dis/enable optional modules and overriding of particular values on the submission command line, configure mappings of file-system paths, and other environment specifics.

Wrapper Scripts

Wrapper scripts are put in place instead of the original Torque scripts, but execute those as well to maintain previously given functionality. There are several of these hook up points, for each phase of the workload management. For a description of tasks carried out during the different phases of job deployments by those wrapper scripts, please refer to Sect. 2.2.

The root level scripts are executed on the bare metal hosts and prepare the virtual guests. They are used by vTorque to prepare and cleanup compute node environments with administrator privileges and to instantiate the guests.

Root Level Wrappers

- prologue
- prologue.parallel
- epilogue
- epilogue.parallel

Another set of wrapper scripts is executed in the user space on bare metal and take care of the actual vm preparations and job start in the virtual guest environment when instantiated by the other root wrapper scripts.

User Level Wrappers

- prologue
- prologue.parallel
- jobscript

Last set of scripts are executed within the virtual guest with administrator privileges. These scripts are hooked into the VM life-cycle by the help of NoCloud metadata and mirror bare metal functionality of Torque in the virtual environment. The scripts are skeletons with logging where additional functionality can be added.

VM Root Level Scripts

- prologue
- prologue.parallel
- epilogue
- epilogue.parallel
- epilogue.precancel

Template Files

There are two categories of templates provided by vTorque. Script templates with placeholders populated during different phases of a batch job's life-cycle. And VM templates for domain definitions and contextualization of virtual guests during instantiation. Both set of files can be modified by administrators according to their infrastructure and user needs.

User level scripts, populated at the time of job submission, are used to prepare the virtual guest instantiation.

User Level Hooks

- vmPrologue.parallel.sh
- vmPrologue.sh
- jobwrapper.sh

While the VM template files are defining the virtual guest machine properties and also customize the virtual guest environment during boot, such as user accounts and installation of security related updates. Placeholders are populated with global configuration values.

VM Template Files

- domain definitions
- domain fragments
- metadata files

Configuration File

vTorque's global configuration file enables administrators to define, besides default resources for virtual job environments, file-system paths mounted into the guests, networking, and mandatory software packages, also to dis- and enable particular features, such as optional components for virtual I/O optimization or workload monitoring. For a full reference, see the configuration file *config.sh*, there is for each parameter a header comment https://github.com/mikelangelo-project/vTorque/blob/master/src/common/config.sh.

2.4   Command Line Interfaces

vTorque introduces two new command line interfaces (cli). One for end users to submit jobs named *vsub* transparently to a virtualized compute environment. The other cli is for administrators to manage virtual guest images more conveniently and to provide users an overview about available ones, and is named *vmgr*, similar to Torque's queue manager *qmgr*.

vsub

The cli *vsub* is the counterpart of Torque's *qsub* cli [1], compatible with all Torque command line arguments. Torque command line options are passed on without modifications, but sanity checks if, for example, the vCPU count exceeds a provided bare metal resource request. In addition to full *qsub* command line compatibility[2], it provides numerous virtual resource definition capabilities. Such as an user level prologue for the virtual environment, the amount of virtual cores and memory per guest, NUMA pinning map, guest image to use and more.

```
usage: qsub [-a date\_time] [-A account_string]
    [-b secs] [-c [ none | { enabled | periodic |
    shutdown | depth=<int> | dir=<path> |
    interval=<minutes>}... ]
[-C directive_prefix] [-d path] [-D path]
[-e path] [-h] [-I] [-j oe|eo|n] [-k {oe}]
[-K <kill delay seconds>] [-l resource_list]
[-m n|{abe}] [-M user_list] [-N jobname] [-o path]
[-p priority] [-P proxy_user [-J <jobid>]] [-q queue]
[-r y|n] [-S path] [-t number_to_submit]
[-T type]   [-u user_list]
[-w] path [-W additional_attributes]
[-v variable_list] [-V ] [-x] [-X] [-z]

usage:
vsub [<qsub_arguments>] [-gf] -vm <vm_parameters>
    [script]
```

Table 1 vsub arguments[a]

Argument	Default value	Description
img	Any *.img/*.qcow2 file	VM image file for the job execution
ram	In K/M/G/T	Amount of memory per guest, i.e. 15G
distro	debian/ubuntu/redhat/centos/osv	Distro of the image, i.e. debian, redhat, osv
arch	Refer to KVM docs, please	CPU architecture, must match compute nodes
vcpus	Positive number	Amount of vCPU assigned to each guest
vcpu_pinning	true/false/<pinning_file>	Use vCPU pinning or not
vms_per_node	Positive number	Amount of VMs per allocated physical node
vm_prologue	An executable file	Optional user prologue script run in standard Linux guests
vm_epilogue	An executable file	Optional user epilogue script run in standard Linux guests
vrdma	Enabled	True
UNCLOT	Enabled	True
UNCLOT_shmem	In K/M/G/T	Amount of memory per guest for ivshmem, i.e. 1024M
iocm	Enabled	True
iocm_min_cores	Positive number	Define minimum amount of dedicated IOcm cores
iocm_max_cores	Positive number	Define maximum amount of dedicated IOcm cores
fs_type	sharedfs/ramdisk	File-system type, shared fs or ram disk
Disk	Any *.img/*.qcow2 file	Optional persistent disk, mounted at the first VM (rank 0)

[a]Complete list of vsub arguments, extending Torque's qsub command line interface

Please refer to Table 1 for an overview and description of *vsub*'s cli extensions (<vm_parameters>). For all settings there are defaults defined (by administrators) in the global configuration file *config.sh*. The additional argument *-gf* prevents job submission after first set of files has been generated, useful for debugging purposes only.

Additionally, environment variables can be set to override the log level or print a live log on the console, useful for debuggin purposes only. Please refer to Log4Bsh's readme [11] for the full documentation.

vmgr

Second vTorque's cli name *vmgr* provides to administrators simple virtual guest image management capabilities. The name is choses as Torque's queue manager is called qmgr [3]. The *vmgr* cli tool copies under the administrator's user virtual guest images to the image pool directory and allows to store guest image descriptions in flat files.

```
usage:
 vmgr show config
 vmgr show images
 vmgr show image <name>
 vmgr show suspended --job=<jobID>  | --user=<userName>
 vmgr add <image> [<description>]
 vmgr update <image> <description>
 vmgr delete image <name>
 vmgr delete suspended --job=<jobID>|--user=<userName>
```

2.5 HPC Infrastructure Abstraction

Each HPC location requires some adaptation of applications. This comprises linking with available libraries, kernel compatibility, file-system paths, CPU and memory optimizations. On an abstract level as it is provided by virtualization, the most common aspects are: a shared home among compute nodes, a fast intermediate storage for computations (workspace), installation paths of libraries and applications, batch system related compute node environment variables, high speed interconnects (e.g. Infiniband/Omnipath), support for further infrastructure services (IP/DNS/NTP).

These data-center infrastructure specifics are abstracted in Clouds, enabling high flexibility and portability. vTorque adapts this approach by the help of NoCloud metadata technology [9] and XML definitions [10] for virtual guests managed by the help of *libvirt*. The metadata file allows administrators to customize virtual guests and adapt them to their environment. vTorque takes care of mapping and adapting batch-system environment variables in the guests virtual job execution environment, such as the (virtual) compute node list.

For compatibility reasons, vTorque relies on these file-system abstractions and it is mandatory for application developer or guest image providers to consider following mount-point mappings:

- user *$HOME* mounted as */home*
- non-persistent workspace for intermediate data as */workspace*
- cluster wide applications installation path (optional) as */opt-hpc*

Customization of virtual environments, by the help of metadata and XML domain definitions, comprise further to pass through particular hardware devices, such as accelerator cards or GPUs from bare metal host to virtual guest access, or to provide networking adapters like high-speed interconnects.

2.6 Optional External Modules

vTorque does not offer a plug-in architecture, however offers integration with a few external modules. As external optimizations and applications differ too much, vTorque does not provide a generic way for administrators to add new ones easily.

Except those do not need to be supported on the *vsub* command line, but can simply be loaded and removed during pro- and epilogue phases.

Today's drawbacks of virtualization are in first place I/O related overhead. Since HPC targets high performance, this virtualization overhead is a crucial aspect for adoption and must justify the costs to be paid for higher flexibility. Within the EU funded H2020 Project MIKELANGELO, besides vTorque also several components addressing virtualization I/O overheads have been developed and integrated with vTorque.

Each of these external I/O optimizations can be dis/enabled by vTorque administrators, a list of hosts can be defined which are capable of the functionality, as well as default settings for vTorque users can be configured.

vRDMA
Huawei's virtualized RDMA (vRDMA) introduces abstraction infiniband cards and provides RDMA capabilities to multiple VMs running on the same host, e.g. each one in its own NUMA domain [6]. For multiple VMs per host a simple pass-through would make the adapter available to exactly one guest, only.

IOCM
Intel's I/O core manager (IOCM) optimizes throughput of virtual I/O by allocating dedicated cores on the bare metal host system, and is implemented as *vhost* kernel module. The amount of dedicated cores can be assigned as a static count or dynamic range. It has however the requirement of several kernel configuration options to be set and modified, besides being available for Linux Kernel version 3.18 [4].

UNCLOT
XLAB's development, UNikernel Cross Level cOmmunication opTimisation in short UNCLOT enables virtual guests on the same host to communicate via shared memory (ivshmem), rather than via loop back devices with additional overhead and higher latency [5].

Snap-Telemetry Monitoring
Intel's Snap-Telemtry service allows to measure application performance from host over guest level to application level. It provides a flexible plug-in framework and plenty plug-ins for various performance metrics. Snap-Telemetry can be used with vTorque to analyze overheads, identify bottlenecks and fine-tune application behavior, to optimize resource utilization by consolidating workloads of disjunctive resource consumption characteristics [7].

Guest Operating System Support
vTorque supports as guest operating system standard Linux guests (debian and red hat family). Additionally, it supports OSv a single process, single user context switch free lightweight unikernel operating system [8]. OSv images are usually built for exactly one application and have a small footprint, very short booting times and memory consumption. Additional operating systems to be supported would require modification of vTorque's scripts, NoCloud metadata and domain XML template files, there is no generic plug-in approach.

Acknowledgements Special thanks goes to Justin Cinkelj (justin.cinkelj@xlab.si) and Gregor Berginc (gregor.berginc@xlab.si) from XLAB for their support with OSv and UNCLOT integration. As well as to Shiqing Fan (Shiqing.Fan@huawei.com) from Huawei for his support surrounding the challenging vRDMA integration and Yossi Kuperman1 (YOSSIKU@il.ibm.com) from IBM for help with IOcm kernel patching. Further, to Marcin Spoczynski (marcin.spoczynski@intel.com) from Intel for helping with Snap-Telemetry monitoring integration and Nadav Har'El (nyh@scylladb.com) from ScyllaDB taking care of OSv feature requests and bug reports. Also, special thanks to former colleague Uwe Schilling for system administration and testbed setup.

3   vTorque Deployment

vTorque extends Torque, thus a working Torque installation is needed in first place. Other dependencies exist for the management of virtual guests, besides requirements towards the compute environment infrastructure.

3.1   Requirements

vTorque relies on available virtualization capabilities known from Clouds, provided by Qemu and KVM. Additionally, there are infrastructure services required:

- Shared filesystem, for example NFS or Lustre
- DHCP server for VM IPs including a sufficient range

Following software packages are required to be installed on the submission hosts providing *vsub* and on the compute nodes.:

- coreutils
- net-tools
- openssh-client
- cloud-utils
- bash (>= v4.0)
- qemu-kvm
- libvirt-bin
- numad (optional; for automatic vCPU pinning)
- pdsh (optional; for *setup.sh* script)

On the compute nodes following needs to be ensured:

- disabled SSH known hosts file (since VM ssh server keys are generated during boot)
- command 'arp -an' is executable by users (used to determine IP of VMs)

Further, a shared file-system is required for several directories defined in vTorque's configuration file:

- the image pool dir *VM_IMG_DIR*
- user homes *VM_NFS_HOME*
- a fast workspace for intermediate data *VM_NFS_WS*
- cluster wide software installations *VM_NFS_OPT*

3.2   Installation

vTorque comes along with an installation script (*setup.sh*), which is taking care of the deployment in an automated manner, by the help of parallel SSH (PDSH) used to connect to all nodes. The actual deployment does not comprise many tasks and can also be carried out manually. The following steps outline the deployment carried out by the script in detail:

1. Clone vTorque git repository recursively
2. Server setup:

   (a) copy */lib, /src/** (, */test, /doc*) dirs to your target installation directory
   (b) copy */contrib/99-mikelangelo-hpc_stack.sh* to */etc/profile.d/*
   (c) define network setup for VMs in vTorque's config file

3. Compute node setup:

   (a) rename all prologue and epilogue scripts in */var/spool/torque/mom_priv/* to **.orig*
   (b) copy as next vTorque's wrappers as *pro/epilogue** scripts

4. as last ensure permissions are correct

   (a) owner of target dir should be root
   (b) permissions for *pro/epilogue** wrapper scripts must have chmod 500
   (c) all other files are recommended to be set to chmod 444

3.3   Precautions

There are several properties of vTorque which must be understood for secure operations and assessment of vTorque's overall costs introduced for bare metal jobs. As well as environmental dependencies in order to optimize its overhead for the management of virtual guests.

Security

The most crucial aspect is to prevent users from booting images themselves, and not allow users to make use of uploaded VM images as it may provide root access on VMs. The default setting prevents it and should not be changed, except the impact on user's data privacy is fully understood. Can be useful for development environments with trusted users, but poses a high risk for daily operations. The NoCloud metadata file, used to customize virtual guests during instantiation, is write protected to prevent users from gaining root access, installing vulnerable packages they intend to exploit, or to break their own privacy by accident. The metadata template file should be edited by admins with high caution, as it may have a major impact on the HPC environments security and data privacy. Further, virtual guest must not be accessible to any other user accounts than the ones allowed to connect to the bare-metal node while the allocation is in place, exception are administrators.

Performance
vTorque has overall costs, due to its non-invasive wrapper concept, alongside the virtualization overhead. At submission time, files are generated and written onto a shared directory, introducing first overhead for virtual jobs. The root prologue may start too quickly and if file-system is asynchronous, it is required to wait for those files to appear, which is in case of bare-metal jobs obviously always wasted time. The corresponding configuration parameter for administrators is *NFS_TIMEOUT*. Setting it too low results in failing jobs, setting it too high causes unnecessary costs for each job's life-cycle. The wrapper scripts are written in Bash and therefore introduce, regardless of VM job or bare metal execution, some overhead during deployment of jobs in both cases, bare metal and virtualized ones. In case of VM jobs there is additional overhead, e.g. to copy VM images onto each compute node, generate VM related files and wrappers, instantiate the guests and wait for them to become available. Concerning virtualization overheads, the host operating system needs resources to run, e.g. CPU and memory, and need to be fine-tuned in order to decrease overhead costs for virtualized jobs.

Environment
Since the VMs comprising a job's virtual compute node environment need an IP address, and many VMs can be spawn over the day and in parallel, it is crucial to have a sufficiently sized pool of (internal) IPs configured and available. Crashing VMs (e.g. due to problems on the physical nodes) will not signal re-usability to the DHCP server if not properly terminated by vTorque. Those IPs are blocked until the lease time is over, which must be chosen by administrators accordingly to their environment. Lack of IP will lead to failing VM instantiations, as required networking cannot be provided to guests and time will be wasted until timeouts are hit. Configuration of networkfile-systems can also influence operational overhead costs for vTorque, e.g. async configuration demands root prologue to wait for files to appear until a timeout is reached.

4 Discussion

In summary vTorque introduces capabilities to run workloads in virtualized environments, merges the benefits of Clouds into HPC. However, at additional costs for virtualized compute node overheads. Image copying, preparation of contextualization and instantiation of virtual guests. Under consideration of common security measures for HPC environments, e.g. root access must be restricted.

A high level of flexibility, common in Cloud hosting and compute environments, is gained with manifold advantages for all stakeholders. Data-center owners can serve new user groups with conflicting needs, administrators can reduce the efforts for providing new kernels or customized environments to particular user groups, and so end users requirements can be satisfied in most cases. End users may be able to share a physical node with another workload with different characteristics, e.g. memory intensive and CPU intensive applications may be a good fit reducing costs for job execution.

Further advantages for administrators are given by the introduced abstraction, e.g. updates of the physical environment do not require a rebuild of the compute node images provided to users, changing file-system paths do not require a reconfiguration of all applications. Another added value for developers is, they can ship their applications ready for use, prepackaged with the optimal combination of kernel and libraries. Applications for HPC environment can be developed without the need for anything else than commodity hardware, and those prepackaged applications run out of the box in Cloud as in vTorque HPC environments.

On the other hand the overhead for the VM preparation phase is costly, and too expensive for jobs with a short run-time. Also, costs increase at least linear with the amount of nodes assigned, as onto each node a copy of the base image needs to be staged onto. This may lead to network congestion and negatively impact even other job's performance if no dedicated network is used for VM image provisioning.

The optional I/O modules supported by vTorque, the IOCM, vRDMA, and UNCLOT provide optimizations, but have their limitations, e.g. patch for particular kernel version or Infiniband driver, and also their potential costs need to be considered, e.g. reserved cores for I/O are not available to the guests. Also, their deployment in a batch environment is more complex compared to vTorque.

Prepackaging of applications in a generalized way however cannot take environment specific optimizations into account, such as particular L1/L2/L3 cache size. Also, in HPC environments often optimized compilers are provided, such as Intel Compiler or Cray compiler. The use of commercial applications, debugging and tracing tools, provided in the HPC environment for bare metal compute nodes, can be used in virtual guests as well if binary compatibility given. These applications and libraries are made available to the guest by mounting the global installation directory as '/opt-hpc'. Environment specific license-servers, and generic hardware support are not integrated, yet, and are part of future work.

Conclusion is, vTorque accomplishes its target of introducing Cloud-like virtualization and flexibility into HPC environments with outlined benefits for all involved stakeholders, but at additional costs. Depending on an environment's general workload characteristics, virtualized applications could be consolidated on available physical resources in a isolated way. And so increase the overall utilization in the data-center, to reduce or even compensate virtualization costs.

5 Future Work

Future work surrounding vTorque may focus on supporting mixed resource requests, global spare node capabilities in combination with node health monitoring and auto-migration of running instances, suspend and resume functionality of virtual guests. Suspend and resume capabilities require support on the middleware layer (e.g. MPI) to handle timeouts of suspended and resumed communication between nodes.

On the virtualization layer future work may focus on further HPC infrastructure support and abstraction, such as license-servers and capabilities to pass through any hardware, e.g. accelerator cards. White-listed metadata targets provide users with more customization capabilities of virtual guest images during instantiation, while introducing potentially security risks.

The operating system can be stripped down to the basic functionality required for virtualization, in order to gain more resources for the virtual layer, reduce maintenance and testing efforts for upgrading the compute nodes and at the same time reducing attack vectors on host level.

Guest operating system level can be extended with support for SUSE based Linux derivatives, Docker and LXC containers.

A native upstream implementation would improve security and reduce overhead further, e.g. for flag file polling. Interactive VM jobs are desirable as well for job-script debugging purposes, however require modification of the source code. Since Torque is no longer open-source as of June 2018 PBSPro can be extended instead.

Acknowledgement

vTorque has been developed within project MIKELANGELO—MIcro KErneL virtualizAtioN for hiGh pErformance cLOud and hpc systems, Co-funded by the European Commission under Horizon 2020 Framework Program of the European Union as H2020-ICT-07-2014: Advanced Cloud Infrastructures and Services program.

RIA project no. 645402, from January 2015 to December 2017.

References

1. Adaptive Computing, Inc.: Torque. Documentation. http://docs.adaptivecomputing.com/torque/6-1-2/adminGuide/torque.htm. Cited 01 Jul 2019
2. Adaptive Computing, Inc.: Torque. qsub Documentation. http://docs.adaptivecomputing.com/torque/6-1-2/adminGuide/torque.htm#topics/torque/commands/qsub.htm
3. Adaptive Computing, Inc.: Torque. qmgr Documentation. http://docs.adaptivecomputing.com/torque/6-1-2/adminGuide/torque.htm#topics/torque/commands/qmgr.htm
4. EU Project MIKELANGELO: IOCM. Technology Description. https://www.mikelangelo-project.eu/technology/iocm-io-core-manager/. Cited 01 Jul 2019
5. EU Project MIKELANGELO: UNCLOT. Technology Description. https://www.mikelangelo-project.eu/technology/UNCLOT-unikernel-cross-level-communication-optimisation/. Cited 01 Jul 2019
6. EU Project MIKELANGELO: vRDMA. Technology Description. https://www.mikelangelo-project.eu/technology/iocm-io-core-manager/. Cited 01 Jul 2019
7. EU Project MIKELANGELO: Snap-Telemetry. Technology Description. https://www.mikelangelo-project.eu/technology/full-stack-monitoring/. Cited 01 Jul 2019
8. EU Project MIKELANGELO: Snap-Telemetry. Technology Description. https://www.mikelangelo-project.eu/technology/universal-unikernel-osv/. Cited 01 Jul 2019
9. RedHat: Cloud Datasources. NoCloud. https://cloudinit.readthedocs.io/en/latest/topics/datasources/nocloud.html. Cited 01 Jul 2019
10. LibVirt: Documentation. Domain XML format. https://libvirt.org/formatdomain.html. Cited 01 Jul 2019
11. Struckmann, N.: Log4bsh. Documentation. https://github.com/mikelangelo-project/log4bsh/blob/master/README.md. Cited 01 Jul 2019

Integrating SDN-Enhanced MPI with Job Scheduler to Support Shared Clusters

Keichi Takahashi, Susumu Date, Yasuhiro Watashiba, Yoshiyuki Kido, and Shinji Shimojo

Abstract SDN-enhanced MPI is a framework that integrates the network programmability of Software-Defined Networking (SDN) with Message Passing Interface (MPI). The aim of SDN-enhanced MPI is to improve MPI communication performance by dynamically steering the traffic within the interconnect based on the communication pattern of applications. A major limitation in the current implementation of SDN-enhanced MPI is that multiple jobs cannot be executed concurrently. This paper removes this limitation by integrating SDN-enhanced MPI with the job scheduler of the cluster. Specifically, we have developed a plugin for the job scheduler that collects and reports the job information to the interconnect controller. A preliminary evaluation demonstrated that applications could gain up to 2.56× speedup in communication.

1 Introduction

The demand for high-performance computing (HPC) clusters has been ever-growing. In fact, exascale machines are now on the horizon. To meet the sustained growth of HPC clusters, the high-performance network that interconnects the compute nodes composing a cluster, or *interconnect*, needs to be enhanced to achieve larger scale, higher bandwidth and lower latency. As a result, the interconnect now accounts for a significant portion of total financial cost and power consumption of an HPC cluster [6].

Until today, the established strategy for designing an interconnect has been over-provisioning, where extra bandwidth and routes are provisioned in the interconnect.

K. Takahashi (✉)
Nara Institute of Science and Technology, Ikoma, Nara, Japan
e-mail: keichi@is.naist.jp

S. Date · Y. Watashiba · Y. Kido · S. Shimojo
Cybermedia Center, Osaka University, Ibaraki, Osaka, Japan
e-mail: date@cmc.osaka-u.ac.jp; watashiba-y@cmc.osaka-u.ac.jp; kido@cmc.osaka-u.ac.jp; shimojo@cmc.osaka-u.ac.jp

© Springer Nature Switzerland AG 2020
M. M. Resch et al. (eds.), *Sustained Simulation Performance 2018 and 2019*,
https://doi.org/10.1007/978-3-030-39181-2_13

The reason for adopting this design strategy is twofold. First, networking hardware used in conventional interconnects do not allow administrators to change their configurations on-the-fly. Therefore, the interconnect needs to be provisioned with abundant bandwidth and routes so that a specific application does not suffer from degradation of communication performance compared to others. Second, a production HPC cluster is usually shared among many users where each user runs different applications. Therefore, tailoring an interconnect to a single application with a specific communication pattern is unrealistic.

However, an over-subscribed design is becoming increasingly challenging to implement due to its rapidly rising financial cost and power consumption. Meanwhile, the long-standing assumption that interconnects are static and cannot be reconfigured no longer holds with the recent emergence of networking technologies that introduce network programmability. A prominent example of such networking technology is Software-Defined Networking (SDN), which is a novel networking architecture that allows administrators to dynamically and flexibly control network devices like a software. We believe that dynamically controlling the traffic in the interconnect in response to the communication pattern of an application can alleviate traffic congestion in the interconnect and remove the need for excessive over-provisioning.

Based on this idea, we have been developing *SDN-enhanced MPI* [3], a framework that integrates Software-Defined Networking (SDN) into Message Passing Interface (MPI) [7]. The goal of this framework is to mitigate congestion in the interconnect and improve MPI communication performance by dynamically steering the traffic in the interconnect based on the communication pattern of application. To this end, we have demonstrated that individual MPI collectives are accelerated by utilizing SDN [2, 9]. Furthermore, we have designed and implemented a mechanism to synchronize the progress of an application and the reconfiguration of the interconnect [10, 12]. Furthermore, we have developed a toolset for facilitating the development of SDN-enhanced MPI that consists of a profiler to extract communication pattern from applications and an interconnect simulator to predict the traffic generated in an interconnect [11].

Although our work so far has successfully demonstrated that MPI applications can be accelerated through the application of SDN, a limitation has still remained towards the deployment of our framework on production clusters: our current SDN-enhanced MPI framework does not support the concurrent execution of multiple applications on a cluster. Since production clusters are usually shared by many users, it is vital that our framework supports clusters running multiple jobs for the practicality.

In this paper, we aim to lift this limitation by integrating our framework into the job scheduler of a cluster. The rest of this paper is organized as follows. Section 2 gives an overview of the key technologies behind our proposal and clarifies the challenges to be tackled. Section 3 presents the architecture of the proposed framework. Section 4 shows the preliminary evaluation result of the proposed framework. Section 5 concludes this paper and discusses future work.

2   Background

2.1   Software-Defined Networking (SDN)

Software-Defined Networking (SDN) [4] is a novel networking architecture that brings programmability into the network and allows users to dynamically and flexibly control the network as if the network was a software. In conventional networking architectures, the *control plane*, which makes the decision on how to handle packets, and the *data plane*, which forwards packets, are tightly coupled together on a single networking device such as a switch. In contrast, SDN separates these two planes into different hardware: the data plane is handled by each networking device, whereas the control plane is handled by a centralized software controller. Administrators are able to dynamically and flexibly control the network by developing a controller that implements their desired network control policy.

The current de facto standard implementation of SDN is OpenFlow [5]. In an OpenFlow network, the data plane is handled by OpenFlow switches, and the control plane is handled by an OpenFlow controller. Each OpenFlow switch holds a flow table, a collection of flow entries. A flow entry defines what action to perform on what type of packets. The OpenFlow controller controls the traffic in the network by installing and updating the flow table on each switch in the network. This paper assumes that the interconnect of the cluster is built using OpenFlow switches.

2.2   Job Scheduler

As mentioned in Sect. 1, a production HPC cluster is a shared environment. Therefore, the computing resources within the cluster need to be efficiently managed and distributed among multiple users.

To achieve this goal, the administrator of a cluster usually deploys a *job scheduler*. A job scheduler is a system that manages the computing resources such as CPUs, GPUs and memories in a cluster. The job scheduler accepts *job* submissions from users, which are requests to run applications on the cluster accompanied by a set of resource requests to run the application. The job scheduler allocates computing resources in the cluster and launches the application on the allocated computing resources. If there are insufficient available resources in the cluster, the job will be queued for later execution. Job schedulers used in production HPC clusters include Slurm [13], PBS Professional,[1] Torque[2] and Grid Engine.[3] In this

---

[1] https://www.pbspro.org/.

[2] https://www.adaptivecomputing.com/products/torque/.

[3] http://www.univa.com/products/.

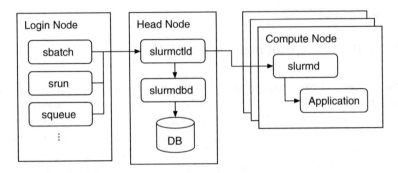

Fig. 1 Architecture of Slurm

paper, Slurm is assumed to be deployed on the cluster since it is one of the most widely adopted open-source job schedulers.

Figure 1 illustrates the architecture of Slurm. Slurm employs a *master-worker* architecture like many other job schedulers. The master, *slurmctld*, oversees the status of every compute node in the cluster. Each compute node runs a worker, *slurmd*, that monitors the status of the compute node, communicates with slurmctld and launches user applications when requested by slurmctld. Slurmctld receives job submissions from users, allocates computing resources to the job and dispatches the job by instructing the workers. Optionally, *slurmdbd* is used for logging and accounting purposes. Users interact with slurmctld through utilities such as *sbatch* (submits a job), *srun* (submits an interactive job) and *squeue* (list queued jobs).

2.3 Challenges

Even though a job scheduler is useful for both administrators and users of a cluster, it causes a major challenge in SDN-enhanced MPI. The interconnect controller needs to know several pieces of information about jobs to compute and install a routing to the interconnect. This information includes:

- When a job starts and finishes
- Which compute nodes are assigned to a job
- How MPI processes are distributed across the allocated compute nodes

However, this job information is only known to the job scheduler and user application. Therefore, the job information needs to be obtained from the job scheduler or user application and then conveyed to the interconnect manager. In addition, the mechanism should be transparent from the users so that they do not need to invoke a special program from their application or link a library to their application.

3 Proposal

This section first briefly reviews the overall architecture of the proposed framework. Subsequently, individual components of the framework are described in detail.

3.1 Overview

The basic idea behind the proposed framework is to integrate the SDN-enhanced MPI framework with the job scheduler so that the reconfiguration of the interconnect can be performed in accordance with the execution of jobs. Figure 2 illustrates the overall architecture of the proposed SDN-enhanced MPI framework. The proposed framework mainly consists of three components: (1) interconnect manager, (2) scheduler plugin, and (3) OpenFlow controller. The interconnect manager is responsible for computing optimized routes for each job. The scheduler plugin is responsible for collecting and submitting the job information to the interconnect manager when a job starts or finishes. The OpenFlow controller is responsible for communicating with the OpenFlow switches and installing the routing generated by

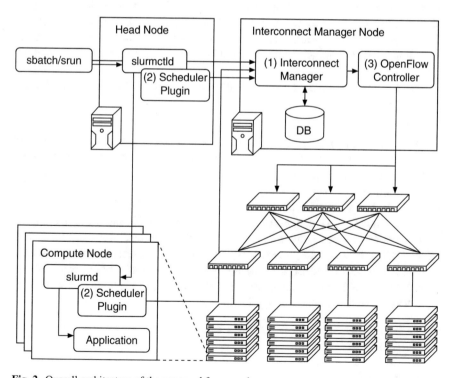

Fig. 2 Overall architecture of the proposed framework

the interconnect manager. We reuse a generic OpenFlow controller provided by the Ryu[4] OpenFlow framework that provides a suite of REST APIs to install, query, update and remove flows.

The scheduler plugin and the interconnect manager communicate with one another using RPCs. Specifically, gRPC,[5] which is an RPC framework built on top of HTTP/2, is used. The main reason behind this design choice is because gRPC can automatically generate server and client codes from an interface definition of remote procedures. Therefore, using gRPC saves much development effort than implementing our own protocol upon raw TCP/IP.

3.2 Scheduler Plugin

The scheduler plugin is responsible for collecting and sending information about a job every time a job starts or finishes. As described earlier, the scheduler decides when to run each job. Furthermore, the node allocation and process mapping are unknown until the job starts. Therefore, a mechanism is needed to notify the interconnect manager of this information.

For this reason, we utilize a built-in plugin mechanism of Slurm, which is called Slurm Plug-in Architecture for Node and job Control (SPANK). SPANK allows developers to easily customize the job startup and cleanup routines of Slurm. SPANK plugins are not linked with Slurm itself and can be loaded during runtime. In addition, SPANK allows developers to add new options to the job script and job submission commands.

Our SPANK plugin is loaded by all Slurm components and performs the following operations:

- **sbatch/srun**: When a job is submitted by a user via sbatch or srun, our plugin sends the ID and name of the job, uid of the submitter, number of processes and communication pattern of the application to the interconnect manager. Currently the user needs to manually specify the communication pattern in the job script as shown in Listing 1 (line 7).
- **slurmd**: When slurmd is about to launch an application on a compute node, our plugin sends the ID and name of the job, compute node ID, and MPI rank number to the interconnect manager. This is used by the interconnect manager to obtain the node allocation and process mapping. After this information is sent over to the interconnect manager, the plugin blocks until the routing are computed and installed to the interconnect. Once the routing is setup, the plugin returns control to Slurm. After that, the user application starts.

---

[4]https://osrg.github.io/ryu/.

[5]https://grpc.io/.

```
1  #!/bin/bash
2  #
3  #SBATCH --job-name=cg-benchmark
4  #SBATCH --ntasks=128
5  #SBATCH --time=01:00:00
6  #
7  #SBATCH --comm-pattern=cg-c-128
```

Listing 1 An example of a job script

When a job finishes, the same information is sent over to the interconnect manager and blocks until the routings are uninstalled from the interconnect. After that, the rest of the cleanup is completed.

All of the above operations are performed transparently from the user application. In other words, the user application does not need to be modified nor a special program needs to be invoked.

3.3 Interconnect Manager

The primary purpose of the interconnect manager is to compute and install optimal routing for each job.

To compute the optimal routing for a job, the interconnect manager needs to know how the MPI processes constituting a job is laid out on the cluster. This information is received from the scheduler plugin integrated into the Slurm scheduler. Furthermore, the communication pattern of a job is also received from the plugin if the user specifies the communication pattern. This job information is persisted on an external database (currently SQLite[6] is used) for fault-tolerance.

Once the scheduler has received the job information, the interconnect manager computes the optimal routing for the job. The actual routing algorithm itself is pluggable and can be swapped out. The computed routing is then installed through the OpenFlow controller to each switch in the interconnect.

4 Evaluation

In this section, we conduct a preliminary evaluation to assess if applications can be accelerated using our proposed framework.

---

[6]https://www.sqlite.org/index.html.

4.1 Evaluation Environment

The evaluation experiment was conducted on a small-scale cluster composed of 20 compute nodes connected through a two-tier fat-tree interconnect (Fig. 3). Each compute node is equipped with two quad-core Intel Xeon E5520 CPUs. In total, there are 160 CPU cores in the cluster. A single NEC PF5240 OpenFlow switch is divided into six virtual switches to compose a fat-tree topology. The switches are controlled by the interconnect manager using the OpenFlow 1.0 protocol. D-mod-K routing [8] was chosen as the representative example of a conventional routing algorithm. D-mod-K routing statically distributes traffic flows across multiple paths in the interconnect based on the destination of a flow.

We executed a set of benchmarks on a cluster and compared the communication time of each benchmark with and without our framework. Table 1 shows the list of communication benchmarks used in the evaluation. CG and FT are taken from the NAS parallel benchmark suite [1]. Stencil2D, Stencil3D, Butterfly and SpMV were developed by us. All benchmarks were executed using 160 processes. In other words, we ran a single process for every CPU core in the cluster.

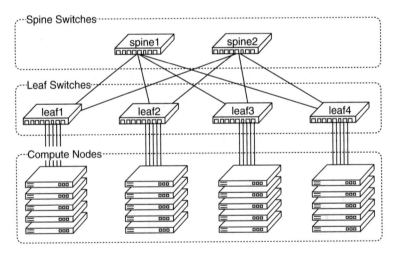

Fig. 3 Cluster used for evaluation

Table 1 Benchmarks used in the evaluation

Name	Description
CG	Solves a sparse linear system using the Conjugate Gradient method
FT	Solves a partial differential equation using FFT and IFFT
Stencil2D	A two-dimensional stencil kernel
Stencil3D	A three-dimensional stencil kernel
Butterfly	An Allreduce kernel commonly seen in deep learning
SpMV	A sparse matrix-vector multiplication kernel

4.2 Evaluation Result

Figure 4 shows the relative speedup of MPI communication when using our framework. CG, Butterfly, and SpMV achieved 1.29×, 1.18×, and 2.56× speedup, respectively. In contrast, FT, Stencil2D, and Stencil3D did not exhibit a clear performance gain by using our framework.

We believe that this trend can be explained for the following reasons. FT performs all-to-all communication between processes. This complete lack of locality makes it challenging to efficiently map the communication pattern into the interconnect. In contrast, Stencil2D and Stencil3D perform nearest neighbor communication in a two-dimensional or three-dimensional process grid. As a result, most of the communication happens within a compute node or within a leaf switch, which makes the difference in routing algorithms irrelevant.

The three benchmarks that benefited from our framework (CG, Butterfly and SpMV) require both local and remote communication. Therefore, there is a chance to improve the load balancing of flows by considering the communication pattern of benchmarks.

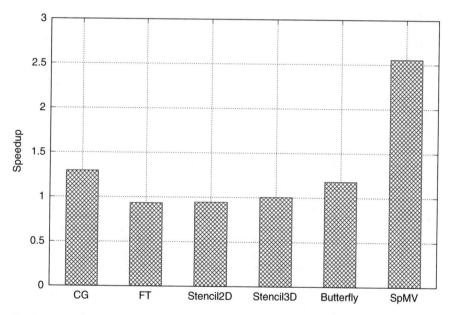

Fig. 4 Benchmark results using miniapps

5 Conclusion

SDN-enhanced MPI aims to improve MPI communication performance by dynamically steering the traffic in the interconnect based on the communication pattern of applications. This paper tackled the limitation of SDN-enhanced MPI that multiple jobs cannot be executed concurrently. Specifically, we integrated SDN-enhanced MPI with the job scheduler. A preliminary evaluation conducted using benchmarks demonstrated that our framework achieves up to $2.56\times$ speedup in communication. As a future work, we would like to evaluate our framework using a large job set containing diverse real-world applications.

Acknowledgement This work was supported by JSPS KAKENHI Grant Number JP17K00168 and JP26330145.

References

1. Bailey, D., Barszcz, E., Barton, J., Browning, D., Carter, R., Dagum, L., Fatoohi, R., Frederickson, P., Lasinski, T., Schreiber, R., Simon, H., Venkatakrishnan, V., Weeratunga, S.: The NAS parallel benchmarks. Inter. J. High Perform. Comput. Appl. **5**(3), 63–73 (1991). https://doi.org/10.1177/109434209100500306

2. Dashdavaa, K., Date, S., Yamanaka, H., Kawai, E., Watashiba, Y., Ichikawa, K., Abe, H., Shimojo, S.: Architecture of a high-speed MPI-Bcast leveraging software-defined network. In: European Conference on Parallel Processing. Lecture Notes in Computer Science, vol. 8374, pp. 885–894. Springer, Berlin (2014). https://doi.org/10.1007/978-3-642-54420-0_86

3. Date, S., Abe, H., Khureltulga, D., Takahashi, K., Kido, Y., Watashiba, Y., U-chupala, P., Ichikawa, K., Yamanaka, H., Kawai, E., Shimojo, S.: SDN-accelerated HPC Infrastructure for scientific research. Inter. J. Inform. Tech. **22**(1), pp. 1–30 (2016)

4. Jamalian, S., Rajaei, H.: Data-Intensive HPC tasks scheduling with SDN to enable HPC-as-a-service. In: 2015 IEEE 8th International Conference on Cloud Computing (CLOUD 2015), pp. 596–603 (2015). https://doi.org/10.1109/CLOUD.2015.85

5. McKeown, N., Anderson, T., Balakrishnan, H., Parulkar, G., Peterson, L., Rexford, J., Shenker, S., Turner, J.: OpenFlow: enabling innovation in campus networks. Comput. Commun. Rev. **38**(2), 69–74 (2008). https://doi.org/10.1145/1355734.1355746

6. Michelogiannakis, G., Ibrahim, K.Z., Shalf, J., Wilke, J.J., Knight, S., Kenny, J.P.: APHiD: hierarchical task placement to enable a tapered fat tree topology for lower power and cost in HPC networks. In: 17th International Symposium on Cluster, Cloud and Grid Computing (CCGrid 2017), pp. 228–237 (2017). https://doi.org/10.1109/CCGRID.2017.33

7. MPI Forum: MPI: A Message-Passing Interface Standard (2012). https://www.mpi-forum.org/docs/mpi-3.0/mpi30-report.pdf

8. Rodriguez, G., Minkenberg, C., Beivide, R., Luijten, R.P., Labarta, J., Valero, M.: Oblivious routing schemes in extended generalized fat tree networks. In: 2009 International Conference on Cluster Computing (CLUSTER 2009), pp. 1–8 (2009). https://doi.org/10.1109/CLUSTR.2009.5289145

9. Takahashi, K., Khureltulga, D., Watashiba, Y., Kido, Y., Date, S., Shimojo, S.: Performance evaluation of SDN-enhanced MPI allreduce on a cluster system with fat-tree interconnect. In: 2014 International Conference on High Performance Computing & Simulation (HPCS 2014), pp. 784–792 (2014). https://doi.org/10.1109/HPCSim.2014.6903768

10. Takahashi, K., Khureltulga, D., Munkhdorj, B., Kido, Y., Date, S., Yamanaka, H., Kawai, E., Shimojo, S.: Concept and design of SDN-enhanced MPI framework. In: Third European Workshop on Software Defined Networks (EWSDN 2015), pp. 109–110 (2015). https://doi.org/10.1109/EWSDN.2015.72
11. Takahashi, K., Date, S., Khureltulga, D., Kido, Y., Shimojo, S.: PFAnalyzer: a toolset for analyzing application-aware dynamic interconnects. In: 2017 International Conference on Cluster Computing (CLUSTER 2017), pp. 789–796 (2017). https://doi.org/10.1109/CLUSTER.2017.18
12. Takahashi, K., Date, S., Khureltulga, D., Kido, Y., Yamanaka, H., Kawai, E., Shimojo, S.: UnisonFlow: a software-defined coordination mechanism for message-passing communication and computation. IEEE Access 6, 23372–23382 (2018). https://doi.org/10.1109/ACCESS.2018.2829532
13. Yoo, A.B., Jette, M.A., Grondona, M.: SLURM: simple Linux utility for resource management. In: Job Scheduling Strategies for Parallel Processing, pp. 44–60. Springer, Berlin (2003). https://doi.org/10.1007/10968987_3

Part IV
Load Balancing Problems
in HPC Simulations

A Method to Reduce Load Imbalances in Simulations of Solidification Processes with Free Surface 3D

Johannes Müller, Philipp Offenhäuser, Martin Reitzle, and Bernhard Weigand

Abstract The present work introduces a first step towards reducing load imbalances in simulations of phase change processes with Free Surface 3D (FS3D). FS3D is a program for the Direct Numerical Simulation (DNS) of multiphase flows. It is able to simulate complex deformations of interfaces between phases by means of a Volume-of-Fluid (VOF) method. The work is focused on the model for the phase transition process from supercooled water to hexagonal ice. In order to investigate complex phenomena with a high computational cost FS3D uses the computing power of the supercomputer Cray XC-40 at the High Performance Computing Center (HLRS). During the calculations, the computational costs for elements that contain both the solid and the fluid phase is higher than for elements which contain only one of the phases. If the computational domain is decomposed into equal parts, the workload is inhomogeneously distributed among the cores. The presented method is able to distribute the workload more homogeneously among the cores and, therefore, enables an efficient use of the computational resources. Elements with higher computational costs are identified by the Volume-of-Fluid (VOF) method. Consequently these elements are associated with a higher computational load in form of a weight. This information is passed to a recursive bisection algorithm which performs the domain decomposition. The recursive bisection of the computational domain considers the existing data structure of FS3D and provides contiguous arrays. To realize the process communication, a nearest neighbour communication was implemented with non-blocking Message Passing Interface (MPI) routines. The diagonal elements are transported via a communication sequence in order to avoid communication of small amounts of data which minimizes the communication overhead.

J. Müller (✉) · M. Reitzle · B. Weigand
Institute of Aerospace Thermodynamics, University of Stuttgart, Stuttgart, Germany
e-mail: johannes.mueller@itlr.uni-stuttgart.de

P. Offenhäuser
High Performance Computing Center Stuttgart, Stuttgart, Germany
e-mail: offenhaeuser@hlrs.de

© Springer Nature Switzerland AG 2020
M. M. Resch et al. (eds.), *Sustained Simulation Performance 2018 and 2019*,
https://doi.org/10.1007/978-3-030-39181-2_14

1   Introduction

Solidification processes play an important role in nature and many industrial applications. This is why they have been studied intensively in recent decades. One example is the naturally occurring phase transition of supercooled water droplets (metastable condition with temperatures well below 0 °C) to ice in clouds [1]. The strong coupling of the different length scales further complicates the development of models: pressure and temperature distributions in clouds, as a consequence of the macroscopical motion, influence the nucleation events (the spontaneous aggregation of water molecules) on the nanoscale. This nucleus can then grow into its supercooled melt while releasing latent heat which significantly influences the temperature inside the droplet and hence impacts on the macroscopical motion [2]. Describing the process as a whole is thus very difficult, if not impossible. Understanding the process on the micro-scale is a first step towards developing models for larger-scale simulations.

The numerical phase change model of water to hexagonal ice developed by Reitzle et al. [3] is shortly summarized in Sect. 2. Here, a model for the anisotropic solidification behaviour of water is coupled to the solution of the energy equation which determines the speed with which the interface moves into its undercooled melt and the amount of latent heat that is released at the interface. All models and algorithms are implemented into the in-house software package Free Surface 3D (FS3D). For details on the numerical schemes and capabilities of the code, the reader is referred to [4]. Different numerical methods exist to reproduce the interface motion. For a summary of different numerical and analytical approaches used to understand the phase transition process from liquid to solid the reader is referred to [3]. Of course a high spatial and temporal resolution is necessary to correctly describe the interface, its deformation, and possible branching. High computing power is required to meet the requirements for accuracy and running time. Current supercomputers generate there computing power by running hundreds of thousands of cores in parallel. To make use of this computing power an efficient parallelisation strategy is needed. The standard parallelisation strategy in FS3D is a symmetric decomposition of the computational domain in equal parts. The computational costs for elements that contain both the solid and the liquid phase is higher than for elements which contain only one of the phases. This is due to the geometrical reconstruction of the sharp interface and additional calculations regarding phase change. The symmetric decomposition did not take account of the different computational cost and hence load imbalances occurred. Therefore, we present a new parallelisation strategy, based on a load balancing by recursive domain decomposition in Sect. 3.3. This idea leads to different numbers of neighbouring domains depending on the distribution of the interface in the computational domain. The realization of the necessary process communication is addressed in Sect. 3.4.

The symmetric and new, weighted decomposition are compared for different test cases in Sect. 4.2.1 where we discuss process communication. Finally, the results are summarised in Sect. 5.

2   Numerical Model for the Solidification Process

The basis for the work presented in this manuscript is the in-house software package Free Surface 3D (FS3D). It allows Direct Numerical Simulations (DNS) of multiphase flows by solving the incompressible Navier–Stokes equations in a Finite-Volume formulation [5]. Different phases are captured by a Volume-of-Fluid (VOF) method, where a scalar function f describes the volume fraction of the disperse phase. Hence, we require

$$f = \begin{cases} 0 & \text{in the liquid phase,} \\ ]0, 1[& \text{in interfacial elements,} \\ 1 & \text{in the solid phase.} \end{cases} \tag{1}$$

A sharp interface is reconstructed by means of planes (PLIC—piecewise linear interface calculation) in all control volumes containing an interface.

FS3D was originally designed to investigate droplet dynamics problems [5] but was later extended in order to treat more complex physical problems. An overview of the capabilities of FS3D is given in [4] which also include the numerical modelling of phase change. Of special interest for the present work is the phase change of a liquid to its solid state (solidification) for water. Here, anisotropy effects need to be additionally taken into account. In the following, the fundamental equations and corresponding numerical methods are presented. The reader is referred to [3] for more details.

In the absence of any convective transport and for incompressible fluids, where the densities of the liquid and solid are equal, the governing equation is conservation of energy which in temperature form reads

$$\frac{\partial (\rho c T)}{\partial t} = \nabla \cdot (k \nabla T) + \dot{q}''' \quad \text{with} \quad \begin{cases} c = c_{p,l} & \text{liquid} \\ c = c_s & \text{solid.} \end{cases} \tag{2}$$

Here, T is the temperature, ρ the density, c the specific heat, k the thermal conductivity, and $\dot{q}'''$ denotes a volumetric heat source. Separate temperature fields are introduced for each phase which are coupled at the interface via the Gibbs-Thomson temperature T_Γ and a jump condition that ensures conservation of total energy across the interface. The Gibbs-Thomson temperature in principal describes the melting point depression at the interface due to curvature effects [6]

$$T_\Gamma = T_m \left(1 - \frac{\sigma_0 H_\gamma}{\rho \Delta h^{sl}} \right), \tag{3}$$

where T_m is the melting temperature, σ_0 a reference surface energy density, Δh^{sl} the latent heat of solidification, and H_γ the anisotropic mean curvature. The latter contains a model for the anisotropic behaviour of the solidifying substance. In the present case this is a hexagonal anisotropy in the basal plane. Information about this model and its mathematical formulation can be found in [3]. During the evaluation of H_γ the surface needs to be reconstructed locally as a graph, which is done by means of a height-function technique [7].

The above mentioned jump condition is commonly known as Stefan condition and can be expressed as

$$V_\Gamma \rho \left(\Delta h^{sl} - \left(c_{p,l} - c_s \right) \left(T_m - T_\Gamma \right) \right) = -k_l \nabla T_l \cdot \mathbf{n}_\Gamma + k_s \nabla T_s \cdot \mathbf{n}_\Gamma, \qquad (4)$$

where V_Γ represents the movement of the interface due to phase change in the normal direction $\mathbf{n}_\Gamma$ to the interface. Note that Fourier's law of heat conduction was used to model the heat fluxes. Finally, in order to realize the phase change, an advection equation for the scalar VOF field f can be formulated where the movement of the interface is interpreted as a flux across the control volume boundaries, rather than a volumetric source term

$$\frac{\partial f}{\partial t} + \nabla \cdot (\mathbf{v}_\Gamma f) = f \nabla \cdot \mathbf{v}_\Gamma. \qquad (5)$$

The auxiliary velocity field $\mathbf{v}_\Gamma$ is constructed via a distribution procedure in the region of the interface, which satisfies $V_\Gamma = \mathbf{v}_\Gamma \cdot \mathbf{n}_\Gamma$. A geometric unsplit advection scheme is used to solve Eq. (5) and a Red-Black Gauss-Seidel algorithm in a successive over-relaxation formulation for the solution of the system of linear equations resulting from Eq. (2).

3 Load Balancing Methodology

3.1 Computer Architecture and Parallel Computing

All tests and benchmarks of the new load balancing strategy for FS3D were performed on the supercomputer "Hazel Hen", a Cray XC-40 system, at the High Performance Computing Center Stuttgart. It consists of 7712 compute nodes that are connected via the high-performance Cray Arias interconnect. Each node has 128 GiB of main memory and two sockets, each equipped with a 12-core Intel® Xeon® E5-2680V3 (Haswell) processor with a base frequency of 2.5 GHz. The "Hazel Hen" is a homogeneous, massively parallel system with 185,088 compute cores and round about 987 TiB of main memory. To make use of the compute power of such a massively parallel system, a suitable parallelisation and domain decomposition strategy for the numerical approach is indispensable. Over the decades various methods of domain decomposition were developed and we refer to Teresco et al. [8] for an overview. The basic idea of all decomposition methods is to divide the domain into N_P sub-domains, where N_P is the number of desired compute cores. The resulting portions are then mapped to the hardware.

3.2 Domain Decomposition and Load Imbalance in FS3D

Previously, the computational domain was divided into parts of equal length with perfectly matching sub-domain interfaces. This allows for a very simple and robust formulation of the communication between cores. This situation is depicted in Fig. 1a. However, since the workload is concentrated on the discrete elements around the solid-liquid interface, the workload of the parallel processes will only be well balanced if the interface is homogeneously distributed through the different sub-domains. An imbalance occurs for such a symmetric decomposition if the interface is only located in specific zones of the computational domain and hence only a couple of cores contain domains with interfacial elements. This situation is illustrated in Fig. 1b, where only the sub-domains Γ_4, Γ_6 and Γ_7 contain parts of the interface.

The load balance of a parallelized program can be quantified by the Load Balance Efficiency (LBE). It results from the average computing time $\bar{t}_{calc}$ and the maximum computing time of all processes $max(t_{calc})$

$$LBE = \frac{\bar{t}_{calc}}{max(t_{calc})}. \qquad (6)$$

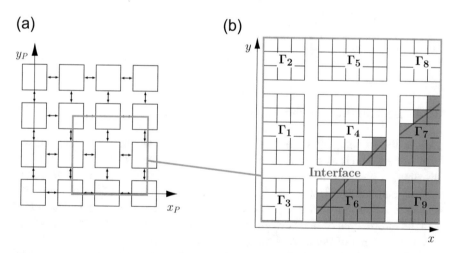

Fig. 1 (a) Symmetric domain decomposition: the computational domain is separated into four sub-domains in each spatial direction. The sub-domains Γ_n and their neighbourhood is depicted as a slice in the $x_P - y_P$-plane. The communication partners are connected by $\leftarrow\rightarrow$. All sub-domains contain the same amount of elements. (b) Enlarged view of a symmetric domain decomposition. The sub-domains Γ_n are shown with their elements. Only the sub-domains Γ_4, Γ_6 and Γ_7 contain interface elements with a higher computational load. The workload is inhomogenously distributed among the sub-domains

For an optimal load balance there exists no load imbalance between the processes, and hence $LBE = 1$. If, however, the load is not distributed equally among the processes we have

$$max(t_{calc}) > \bar{t}_{calc}. \tag{7}$$

For the example in Fig. 1b the calculation times t_{calc} for the sub-domains Γ_4, Γ_6 and Γ_7 are longer and all the other processes have to wait. In order to overcome the degradation of the parallel performance produced by such load imbalances, a new method to decompose the domain is introduced in the following.

3.3 Balanced Domain Decomposition

The main steps of the balanced domain decomposition are listed below and discussed in detail in the next sections. Note that step one and two are not repeated for every timestep but rather after some user-defined number of cycles.

Main steps for the parallel load balancing strategy:
1: Assign a computational cost to each element
2: Create the domain decomposition
3: Calculate the VOF interface operations
4: Exchange the data between the sub-domains

3.3.1 Domain Decomposition by Recursive Bisection

Due to the fact that FS3D uses solely structured, rectangular grids, a recursive bisection method is used in order to minimize cache-misses. Depending on the load distribution given by the sum of the weights, the domain is split iteratively until a given number of sub-domains (cores) is reached. The steps of the bisection are shown in Algorithm 1: first of all, the domain with the highest load is determined and the information of the index coordinates are saved. Afterwards, the size of this domain in terms of number of elements is calculate from the index coordinates. After this, the domain is split into two equal parts, where it is always first divided along the slowest running array index in order to minimize the number of cache-misses. With respect to the Fortran convention the domain is split in the order of: k-index, j-index and i-index. Finally, the index coordinates of the new sub-domain are saved, which describe the extent of a sub-domain, and which are called bisection coordinates. The whole procedure of the bisection is visualized in Fig. 2.

Algorithm 1 Recursive bisection for the domain decomposition

---

recursive bisection algorithm
while (number of subdomains < total number of subdomains)
1: determine domain with highest load
2: duplicate expansion of domain
3: calc number of elements in each direction
4: **if** (number of elements n_k >= number of elements n_j) **then**
5:     bisect number of elements n_k for new domain
6: **else if** (number of elements n_j >= number of elements n_i) **then**
7:     bisect number of elements n_j for new domain
8: **else**
9:     bisect number of elements n_i for new domain
10: **end if**
11: increase number of sub-domains by one
    endwhile

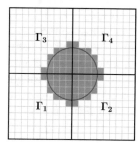 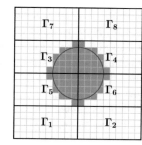 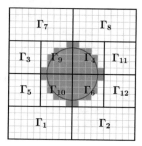

Fig. 2 Domain decomposition by recursive bisection. The computational domain is decomposed into the sub-domains Γ_n. The decomposition from left to right represent steps after three, seven and eleven bisections. Through the bisection, in each case the domain with the highest workload is divided into two sub-domains of equal size

3.3.2 Determination of Load Distribution

To provide the information of the current workload distribution to the bisection algorithm the scalar field $w(i, j, k)$ is introduced. It contains weights which represent an estimation of the workload of the elements. If the weight per cell was known, the computational load for any sub-domain is given by

$$L_\Gamma = \sum_1^N w(i, j, k) \tag{8}$$

where N is the number of elements of the sub-domain Γ. The optimal load per sub-domain is the mean of the whole computational load

$$L^{opt} = L^{mean} = \frac{\sum_1^{N_\Gamma} L_\Gamma}{N_\Gamma}, \tag{9}$$

where N_Γ is the total number of sub-domains. In computational elements containing an interface additional calculations are necessary, e.g. determination of normal vectors, curvature, interfacial velocity, and surface temperature. The position of these elements in the domain is known by the volume fraction field (f−field). These elements are associated with the weight $w_s(i, j, k)$, whereas for all other elements the weight $w_e(i, j, k)$ is used. We set

$$w_s >> w_e. \tag{10}$$

The values of these weights determine the quality of the load balancing. They depend on the total number of elements, the number of surface elements, and the number of processes involved. To determine the optimal weight a priori is therefore not possible. It would have to be determined at runtime of the program. Instead, the weight $w_s(i, j, k)$ is initially estimated. Since for the solution of the energy equation the workload is the same for all elements, we set $w_e(i, j, k) = 1$. Figures 3 and 4 show decompositions for different weights $w_s(i, j, k)$ in a slice through a three-dimensional case where a solid sphere is located in the centre. High weights for $w_s(i, j, k)$ result in a low total number of sub-domains which contain surface elements, and vice versa. Figure 3 shows a domain decomposition if the same weight $w_e(i, j, k) = w_s(i, j, k)$ is used for all elements, and hence, all sub-domains have the same number of elements. Figure 4 shows a decomposition, which occurs when surface elements are assigned a higher computing load $w_e(i, j, k) > w_s(i, j, k)$.

Fig. 3 Domain decomposition with an equal number of elements for all sub-domains. The assigned weigh was for all elements the same $w_s(i, j, k) = w_e(i, j, k)$

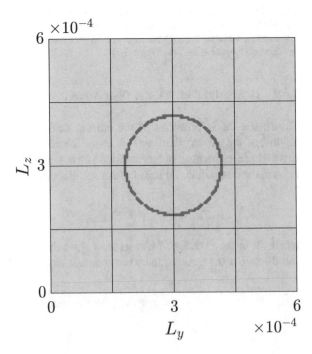

Fig. 4 Domain
decomposition with different
numbers of elements for the
sub-domains. Elements which
contain the surface have been
assigned with a higher weight
$w_s(i, j, k) >> w_e(i, j, k)$

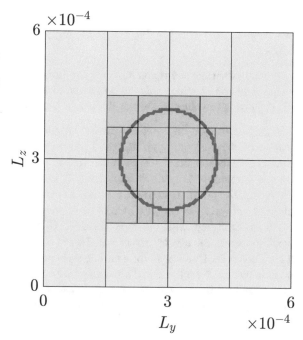

3.4 Process Communication

The main steps to prepare the process communication are listed below and are discussed in detail in the next sections. Note that steps one and two are only repeated if a new domain decomposition was necessary.

Main steps to prepare the process communication
1: Construct the neighbourhood
2: Construction of the buffer coordinates for send and receive buffers
3: Exchange the data between the sub-domains

3.4.1 Constructing the Neighbourhood Relations

For the load balanced domain decomposition, as shown in Fig. 4, the sub-domains have different sizes and can have multiple neighbouring sub-domains per side. All sub-domains are cuboids with sides xl, xr, yb, yt, zb, zf. In order to exchange values between these sub-domains, the processes need the information to which other processes data should be sent. A process is identified by its rank p, the sub-domains are identified by the index n. Since in this work each process handles one sub-domain, $p = n$. For the communication the following information is needed:

- the neighbouring sub-domains of the sub-domain
- the direction on which side the neighbour touches the sub-domain
- the portion of values of the sub-domain to be exchanged

The neighbourhood information of a sub-domain is reconstructed with the information of the bisection coordinates. This is exemplarily shown in Fig. 5 for the side xl of sub-domain T which is oriented in negative x-direction. To determine the sub-domains on the side xl for domain T, in a first step potential neighbours are identified. They fulfill the criteria that they have the same bisection coordinates in x-direction: $ci_{min}(T) = ci_{max}(R)+1$. In the example these are the sub-domains R and F. In a second step, which is depicted in Fig. 6, it is checked whether the potential neighbours are true neighbours or not. This is done by testing if the domains share a common side. Domain F for example does not share a side with domain T and no data has to be exchanged. This is expressed by the criteria if the side is shifted in y-direction: if $cj_{min}(R) \geq cj_{max}(T)$ than F can be excluded as a neighbour. With this two steps, all neighbours of a sub-domain in one direction can be found.

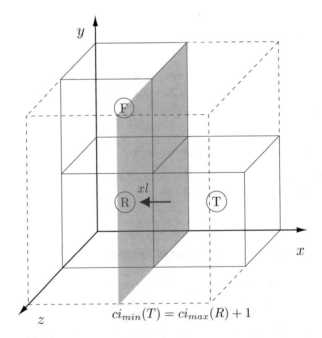

$$ci_{min}(T) = ci_{max}(R) + 1$$

Fig. 5 Construction of potential neighbours for sub-domain T. The right side of the sub-domains R and F are in the same plane as the left side of sub-domain T. Therefore, R and T are potential neighbours

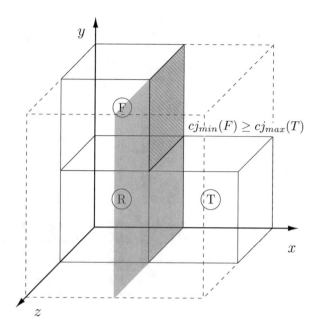

Fig. 6 Construction of true neighbours for sub-domain T. The sub-domains R and F are potential neighbours but the side of F (shaded in blue) is shifted in positive y-direction. Therefore, F can be excluded as a neighbour form T. This is expressed with the criteria for the bisection coordinates $cj_{min}(R) \geq cj_{max}(T)$

With analogous criteria for the other sides of the sub-domain all neighbours of a sub-domain can be determined.

3.4.2   Construct of the Buffer Coordinates for Send and Receive Buffers

If the neighbourhood relationships are known, values can be exchanged between the processes. Since a 27-point stencil is needed for the construction of the normal vectors, the diagonal element depicted in Fig. 7 (green) has to be exchanged as well. However, the exchange of such small amounts of data should be avoided because otherwise the time to start the exchange process outweighs the time for the actual data transfer. The diagonal element is, therefore, not sent directly to the corresponding domain, but transported with a communication sequence described in Sect. 3.4.3. Thus, values to be exchanged come not only from the core partition, but also from the halo element region, since the diagonal elements are transported via this region. The indices of the elements where information needs to be exchanged are called buffer coordinates. They define the set of values which have to be exchanged between the domains. In Fig. 8 this set of values which have to be exchanged from Source to Destination is indicated in red. Because one domain

Fig. 7 Enlarged view of the decomposition. Elements in the core partition are white, elements in the halo region are grey. The exchange of values is realized with a communication sequence. In step ① the elements marked red are send to the right (xr). In step ② the blue elements are send to the top (yt). With this sequence the diagonal element indicated in green is piggybacked and has not to be exchanged directly

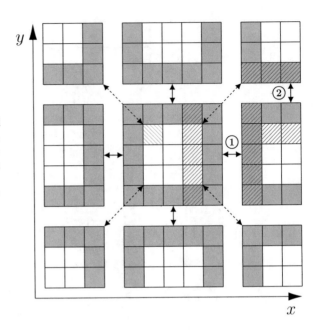

Fig. 8 Representation of values to be exchanged between two domains in positive x-direction. The bisection coordinates of the source are indicated with the index S and respectively with the index D for the Destination. The source is shifted in positive y-direction and has a smaller extension in z-direction than the source. The send coordinates which span the red surface must be calculated

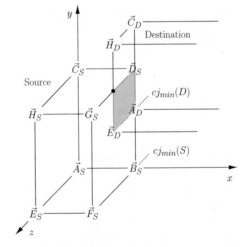

can have multiple neighbours to one side three cases have to be distinguished for the construction of these coordinates. These cases are depicted in Figs. 9, 10, and 11 and are exemplified for the construction of the minimum index coordinate in y-direction:

For case 1 (Fig. 9) the side of the destination domain is shifted in positive y-direction. This means that the bisection coordinate $cj_{min}(Destination)$ of the lower corner of the destination is larger than the bisection coordinate $cj_{min}(Source)$ of the lower corner of the source. The send coordinate sj_{min} results from the bisection

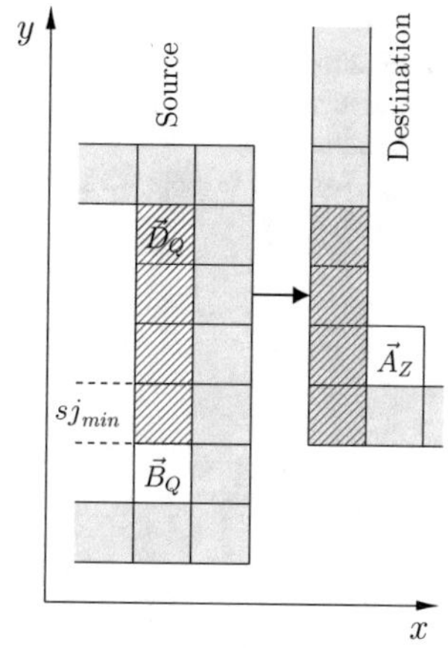

Fig. 9 Case 1: the destination is shifted in positive y-direction. The send coordinate $sj_{min}(Source)$ is calculated with the bisection coordinate $cj_{min}(Destination)$

Fig. 10 Case 2: destination and source are on the same level. The calculation of the send coordinate $sj_{min}(Source)$ is done with the $cj_{min}(Source)$

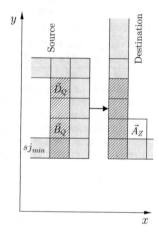

coordinate of the destination $cj_{min}(Destination)$. For the other two cases the send coordinates are calculated by analogous criteria. The maximum index coordinates for the x- and z-direction as well as the receiving coordinates can by calculated according to the same principle.

3.4.3 Communication Routines

The development of an efficient process communication is fundamental for the optimization of the runtime of a program as it can make up a significant part of the total runtime of the program (communication overhead). Decisive for the development of communication between processes is the communication pattern, i.e. which process is exchanging data [9]. From a geometrical perspective the domain decomposition allows for a communication pattern in form of a nearest

Fig. 11 Case 3: the
destination is shifted in
negative y-direction. The
calculation is of
$sj_{min}(Source)$ is done with
$cj_{min}(Source)$

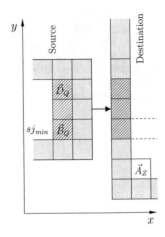

Algorithm 2 Communication routine for the exchange of data in xl-direction. The same sequence follows for the exchange in xr-direction to have a complete exchange of all domains in x-direction. Analogous routines are used for the communication in y- and z-direction

```
    exchange_xl
 1: Read in all neighbours per direction
 2: if neighbours in xl-direction exist then
 3:     for all neighbours in xl-direction do
 4:         fill send buffers (one dimensional array)
 5:         allocate receive buffers (one dimensional array)
 6:         CALL MPI_IRECV (receiving buffer (star_inx)),
                        length of receiving buffer(i), Source(i), ... )
 7:         CALL MPI_ISEND (sendbuffer(start_inx)),
                        (length of send buffer(i), Destination(i), ... )
 8:     end for
 9:     CALL MPI_WAITALL(number of MPI ranks, request)
10:     for all neighbours do
11:         write received buffers back to field
12:     end for
13: end if
```

neighbour communication, since only values have to be exchanged with the direct neighbours. However, an irregular virtual process topology results from the load balanced domain decomposition, since the domains may not be mapped to cores which are next to each other.

As was mentioned above, a communication sequence is introduced where the exchange of small messages is avoided. It consists of three steps: an exchange in x-, y-, and z-direction. The communication to the next spatial direction is not started until the previous communication has been completed (see MPI_WAITALL in Algorithm 2). This sequence allows to transport all diagonal elements in three steps to their final destination.

The implementation of these communication steps was split into three routines. Algorithm 2 exemplarily shows the routine for the exchange in x-direction. The non-blocking MPI calls MPI_ISEND to send values, and MPI_IRECV to receive values were used. With these calls, data can be sent from one process to multiple processes without waiting for one exchange to be finished. From a geometrical point of view, this corresponds to the situation where all values that have to be exchanged in one direction are sent one after the other from one sub-domain to all neighbours in that direction. Although more processes need to communicate with each other, this ensures that the communication overhead increases only slightly.

4 Results

In order to evaluate the new load balancing strategy, numerical test cases were set up, where different load balanced decompositions were compared with the corresponding symmetric one. This was done by timing the execution time for one timestep and a detailed analysis of the calculation and communication time for a routine with VOF calculations.

4.1 Numerical Test Cases

The test cases were inspired by a numerical investigation of the solidification process of an ice seed right before the start of an unstable growth. The initial seed has a diameter of $1.2158 \cdot 10^{-6}$ m. This is the diameter from which instabilities (e.g. dendritic structures) can grow on the surface of the seed [10]. The seed is located in a computational domain with an edge length of five times the diameter in each spatial direction. Since the seed grows with time, a seed with a doubled diameter is initialized in the same computational domain for the second numerical test case. For the third and fourth test case, the spatial resolution was increased to 256 elements in each spatial direction, while keeping the initial diameters as in case one and two. The seed has been initialized for all cases asymmetrical in the computational domain as it is shown for case C128_0.37 in Fig. 12. All test cases are listed in Table 1. The study of the optimal weight per surface element is presented in Sect. 4.2. All test cases have been decomposed always into 16 sub-domains but with different weights per surface element. This results in the different domain decompositions listed in Table 2. In Sect. 4.3 we discuss the parallel efficiency with up to 256 sub-domains by comparing a load balanced decomposition with the symmetric decomposition.

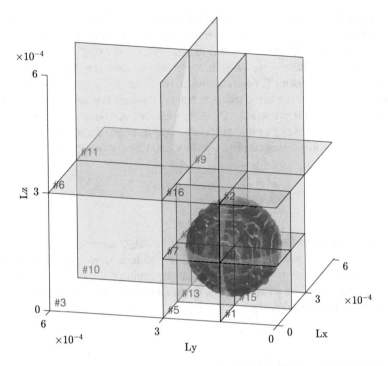

Fig. 12 Visualized domain decomposition LB3 for Case C128_0.37. The computational domain has 128^3 elements with 0.37% of surface elements. Eight processes are used to calculate the domains with surface elements indicated in blue. The different sub-domains are labeled in red. The value of the weight per surface element w_s was 150 times higher than w_e ($w_s = 150$, $w_e = 1$)

Table 1 Numerical test cases with different interface configurations

Case name	Number of surface elements	Percentage of surface elements
C128_0.09	2,097,152	0.09%
C128_0.37	2,097,152	0.37%
C256_0.05	16,777,216	0.05%
C256_0.19	16,777,216	0.19%

4.2   Comparison of Domain Decompositions

In order to derive a criterion for an optimal weight per element, different domain decompositions were compared. To this end, cases C128_0.09 and C128_0.37 have been decomposed into the same number of sub-domains ($N_p = 16$), but with different weights associated to the surface elements. The characteristic parameters of the resulting decompositions can be seen in Table 2. To compare the decompositions, the parameters for a symmetric decomposition are additionally shown in column "SYM". Furthermore, "LB 1" represents a symmetric decomposition since the

Table 2 Characteristics of different domain decompositions for Case128_0.09

	SYM	LB 1	LB 2	LB 3
w_s: weight per surface element	$w_e = w_s$	$w_e < w_s$	$w_e \ll w_s$	$w_e \ll w_s$
Processes with surface elements	2	2	4	8
$\sum f_3$/process C128_0.09	901	901	451	225
$\sum f_3$/process C128_0.37	3897	3897	1949	975
% surface elements/process C128_0.09	0.04%	0.04%	0.02%	0.01%
% surface elements/process C128_0.37	0.19%	0.19%	0.09%	0.05%
Imbalance I	7	7	3	1

The domain is divided in different ways but always into 16 sub-domains. $\sum f_3$ is the total number of surface elements

Fig. 13 Comparison of the computing time for one time step with the load balanced decompositions and the symmetric decomposition. For case C128_0.37 with 0.37% of surface elements the computing time can be reduced by 50%

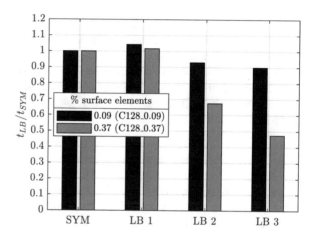

weights were equal for all elements. An increased weight for the surface elements leads to decompositions "LB 2" and "LB 3". This is also exemplarily depicted in Fig. 12 for the decomposition "LB 3" of case C128_0.37. For the decomposition "LB 3" of case C128_0.37 the scalar value of the weight per surface element w_s was 150 times higher than w_e ($w_s = 150$, $w_e = 1$).

4.2.1 Analysis of the Runtime for Different Decompositions

In the following, the runtime is analysed for the different decompositions. To this end, the time that is required for the calculation of one time step t_{LB} is measured and compared to the corresponding time t_{SYM} of a symmetrical decomposition. Figure 13 shows t_{LB} related to t_{SYM} for case C128_0.09 in black and for case C128_0.37 in grey. The total percentage of surface elements is 0.09% and 0.37%, respectively. A comparison of the same decomposition "SYM" and "LB 1", but for "LB1" with the new process communication, shows, that the new implementation is about 0.04% slower. These additional costs are due to the implementation of the process communication. If the decompositions "LB 2" and "LB 3" are compared with the symmetric decomposition, they are faster for both test cases. Note that case C128_0.37 could be calculated in half the time with decomposition "LB 3" compared to the symmetric case. The time for calculating a time step for case 128_0.09 with fewer surface elements can only be reduced by 10%. This means that the larger the ratio of surface elements to the total number of elements, the more pronounced is the load imbalance and a load balanced decomposition becomes effective.

4.2.2 Analysis of the Communication

The percentage of communication time is analysed in detail for one routine which is performing operations on surface elements. These operations are called in the following interface operations. The communication time consists of the time for the actual data transfer and, due to the MPI_WAITALL in the communication routine, also of the duration that processes wait for data from other processes. It can be seen in Fig. 14 that if the load is better distributed among the processes, this waiting time is reduced and hence the share of communication to the total time. This becomes again particularly clear for case C128_0.37 were the share of communication drops to 65%.

4.2.3 Quantification of Load Imbalance

If only routines with interface operations are considered the load imbalance can be quantified in a geometrical way. Following Jofre et al. [11] it can be expressed as the difference between the percentage of the process with the most surface elements and the mean value of surface elements divided by the latter

$$I = \frac{\%\text{surface elements}(P_{max}) - \%\text{surface elements}(P_{mean})}{\%\text{surface elements}(P_{mean})}. \tag{11}$$

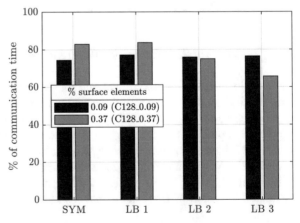

Fig. 14 Percentage of communication time for the different decompositions. The communication time comprises also the time how long the processes have to wait for the slowest process. This is why it is reduced for decompositions which balance the load more equally ("LB 2", "LB 3")

It should be noted that the imbalance increases during the simulation because, as the solidification proceeds, the number of surface elements increases. In contrast to this, the imbalance decreases with a higher spatial resolution, since the total number of elements in the computational area increases cubically and the number of surface elements quadratic. Another aspect that has to be considered is that every routine of the program has a different load balance efficiency. This means that, with a symmetric decomposition, the load imbalance only exists for the routines with interface operations but for routines which have a homogeneous work distribution (e.g. the solution step of the energy equation) there is no load imbalance. Hence, an imbalance is introduced in these routines that were perfectly balanced before. This becomes apparent if the weight for the surface elements is chosen too high. The time saved due to the reduction of the load balance in routines with interface operations is over-compensated by the load imbalance introduced in other routines. This may also lead to a longer overall execution time for one time step.

4.2.4   Detailed Analysis of the Program Run

The situation described in Sect. 4.2.1 is confirmed by a tracing of the program run. Case C128_0.09 is calculated with a symmetric decomposition and a load balanced decomposition. To compare these decompositions an excerpt of the program trace is presented. The tracings excerpts in Figs. 15 and 16 show the 16 processes working on the same part of a routine with interface operations. For a symmetric decomposition the tracing is shown in Fig. 15. Only the two processes 0 and 4 contain surface elements and perform operations on these surface elements (indicated in green), the other processes have to wait (indicated in red) most of the time. In contrast, Fig. 16 shows the balanced decomposition "LB 3" for the same case and a much more evenly distributed workload.

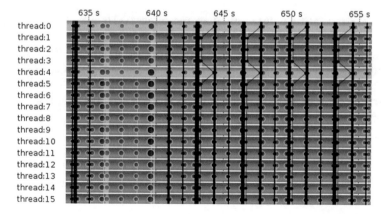

Fig. 15 Tracing of program run for case C128_0.09 with a symmetric decomposition ("SYM")

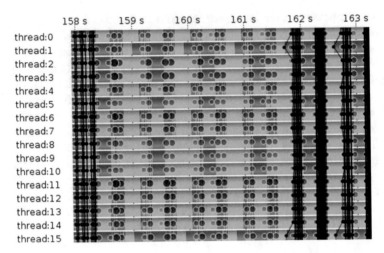

Fig. 16 Tracing of program run for case C128_0.09 with a balanced decomposition ("LB 3")

4.3 Parallel Efficiency

The parallel efficiency for a larger number of processes is examined for the cases C128_0.37 and C256_0.19, since the load imbalance is too small for the two other cases. The ideal speed-up results with the assumption that the execution time with k processes is k-times faster. Since the representation selected here for the achieved speed-up S refers to the base value for the calculation with 16 processes, the ideal speed-up is

$$S_{id} = \frac{P}{16},$$ (12)

where P denotes the number of processes. Figure 17 shows the speed-up for case C128_0.37 with 128^3 elements. The load balancing leads to a significant acceleration for up to 128 cores. However, a maximum of 128 cores can be used for this case, as otherwise the number of elements per process becomes too small. Figure 18 shows the speed-up for case C256_0.19 with 256^3 elements. Since the imbalance is lower for this case, the balanced decomposition leads to lower accelerations.

Fig. 17 Comparison of the speed-ups S for the symmetric decomposition "SYM" and the balanced decomposition "LB 3" for test case C128_0.37 with 128^3 elements and 0.37% surface elements

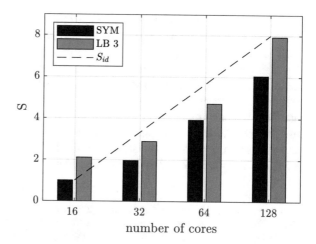

Fig. 18 Comparison of the speed-ups S for the symmetric decomposition "SYM" and the balanced decomposition "LB 3" for test case C256_0.19 with 256^3 elements and 0.37% surface elements

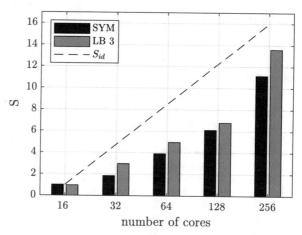

5 Conclusion

In this paper, we have presented a method to reduce load imbalances in simulations of solidification processes with the software package Free Surface 3D (FS3D). Elements which contain an interface have higher computational costs due to operations that are solely performed on the interface. Therefore, these elements are associated with a higher weight compared to non-interface elements. This allows to create a load map that is passed to a recursive bisection algorithm. Therein, the recursive bisection of the computational domain considers the existing data structure of FS3D and provides sub-domains consisting of contiguous arrays. For the data exchange between processes, a nearest neighbour communication was implemented with non-blocking Message Passing Interface (MPI) routines. In order to communicate the diagonal elements, a communication sequence is used to minimizes the communication overhead. The load imbalance occurring in FS3D could be quantified as a function of the percentage of surface elements. Since the load imbalance occurs only in routines which perform calculations on surface elements, different domain decompositions should be used in the future for the different routines of the program. The reduction of the load imbalance was investigated for different domain decompositions and a criterion for an optimal weight per element can now be formulated based on this work. Furthermore, the parallel efficiency of the load balanced domain decomposition was compared to the original domain decomposition. For cases with a large number of interface elements or an asymmetrical distribution of interface elements the simulations can be significantly accelerated by the introduced method.

Acknowledgements This work is funded by the Deutsche Forschungsgemeinschaft (DFG, German Research Foundation)—Project Number 327154368 - SFB 1313, the Collaborative Research Center SFB-TRR75 and was partially executed within the frame of the German SiVeGCS project. The authors gratefully appreciate the access to the high performance computing facility "Hazel Hen" at HLRS, Stuttgart under Grant No. FS3D/11142 and would like to thank the teams of HLRS and Cray for their kind support.

References

1. Sassen, K., Liou, K.N., Kinne, S., Griffin, M.: Highly supercooled cirrus cloud water: confirmation and climatic implications. Science **227**(4685), 411–413 (1985)
2. Pruppacher, H.R., Klett, J.D.: Microphysics of Clouds and Precipitation. Kluwer Academic Publishers, Boston (1997)
3. Reitzle, M., Kieffer-Roth, C., Garcke, H., Weigand, B.: A volume-of-fluid method for three-dimensional hexagonal solidification processes. J. Comput. Phys. **339**(C), 356–369 (2017)
4. Eisenschmidt, K., Ertl, M., Gomaa, H., Kieffer-Roth, C., Meister, C., Rauschenberger, P., Reitzle, M., Schlottke, K., Weigand, B.: Direct numerical simulations for multiphase flows: an overview of the multiphase code FS3D. J. Appl. Math. Comput. **272**(2), 508–517 (2016)
5. Rieber, M., Frohn, A.: A numerical study on the mechanism of splashing. Inter. J. Heat Fluid Flow **20**(5), 455–461 (1999)
6. Alexiades, V., Solomon, A.D.: Mathematical Modeling of Melting and Freezing Processes. Hemisphere Publishing Corporation, New York (1993)
7. Popinet, S.: An accurate adaptive solver for surface-tension-driven interfacial flows. J. Comput. Phys. **228**(16), 5838–5866 (2009)
8. Teresco, J.D., Devine, K., Flaherty, J.E.: Partitioning and dynamic load balancing for the numerical solution of partial differential equations. In: Numerical Solution of Partial Differential Equations on Parallel Computers. Springer, Berlin (2005)
9. Hoefler, T., Traeff, J.L.: Sparse collective operations for MPI. In: Proceedings of the 23rd IEEE International Parallel & Distributed Processing Symposium, HIPS'09 Workshop (2009)
10. Rauschenberger, P., Weigand, B.: A volume-of-fluid method with interface reconstruction for ice growth in supercooled water. J. Comput. Phys. **282**, 98–112 (2015)
11. Jofre, L., Borrell, R., Lehmkuhl, O., Oliva, A.: Parallel load balancing strategy for volume-of-fluid methods on 3-D unstructured meshes. J. Comput. Phys. **282**, 269–288 (2015)

Load Balancing for Immersed Boundaries in Coupled Simulations

Neda Ebrahimi Pour, Verena Krupp, Harald Klimach, and Sabine Roller

Abstract The simulation of engineering problems usually involves different physics and scales that need to be addressed appropriately. A monolithic computation on the finest scale of such complex problems results in overly expensive computations, unfeasible to solve even on today's supercomputing facilities. To facilitate accurate simulations of such problems we suggest a partitioned coupling approach. The strategy is, to decompose the simulation domain according to the physics into subdomains and solve each of them with the best suited numerical approximation. All subdomains are weakly connected to each other at the boundaries (coupling interface), while a coupling approach takes care of the communication and the data-exchange between them. With that we can not only address specific requirements of the physics individually but also reduce the computational cost when compared to the monolithic scenario, where the entire domain is treated with the same equations and numerical scheme. One drawback we face in coupled simulations is the implied load imbalance, which is due to the different treatment of each subdomain and the communication and interpolation between them. These additional loads do not affect the complete domain equally and a balancing strategy within the subdomains is required for efficient computation. To represent geometries in our setups, we employ a penalization method that fits well with high-order discretization schemes, but introduces volumes, where the solution is not of interest. By reducing the scheme order selectively in those regions that are not of interest, the impact of this strategy on the computational effort can be minimized but this introduces another factor of load imbalance. In this work we present observations on these load imbalances and how they can be balanced in the coupled setup, enabling the efficient computation of complex setups as found for example in direct aero-acoustic simulations.

N. Ebrahimi Pour (✉) · V. Krupp · H. Klimach · S. Roller
University of Siegen, Siegen, Germany
e-mail: neda.epour@uni-siegen.de; verena.krupp@uni-siegen.de; harald.klimach@uni-siegen.de; sabine.roller@uni-siegen.de

© Springer Nature Switzerland AG 2020
M. M. Resch et al. (eds.), *Sustained Simulation Performance 2018 and 2019*,
https://doi.org/10.1007/978-3-030-39181-2_15

1 Introduction

Multi-scale and multi-physics problems are of complex nature, and hard to solve numerically, as they often require the interaction of adapted methods for specific parts of the overall setup. Therefore, a proper strategy is needed to simulate these kinds of problems accordingly to not only represent the physics correctly but also to keep the computational effort in a reasonable scope. For that, we consider partitioned coupling, that allows the simulation of multi-scale and multi-physics problems by decomposing the overall setup into spatially separate subdomains and treating each part with the best suited numerical setup. Each subdomain can then be solved individually, taking the physical requirements into account. This allows to solve different equations with individual numerical schemes and the required spatial resolution in each subdomain. With that we can ensure a tailored treatment of the entire simulation domain. Due to the decomposition and the different treatment of each subdomain, a coupling tool is needed to maintain the data-exchange between them. We already showed how efficient this strategy is in terms of computational time, when compared to a monolithic approach, where the entire domain is solved for the same equations with the same numerical scheme and resolution. Even though, this strategy not only reduces the computational time but also provides accurate results [1–3], we face a new bottleneck due to the heterogeneity of the computation. Since each domain is treated differently, load imbalances can be foreseen and have to be addressed in order to use the available computational resources efficiently. There are two problems to consider here. The one is the balancing between subdomains accounting for the individual computational cost in each them. The other is the balancing of unevenly distributed loads within each subdomain. Both are important for large scale simulation runs. When a geometry is involved, the second problem of balancing within each subdomain gets even more important, as we utilize a penalization scheme here and reduce the computational effort where possible. Hence, there are multiple factors influencing the computational cost of individual elements, with the additional costs for the interpolation at the coupling interface and the reduced effort within penalized areas as the most important ones.

In this paper we address this problem for a three-field coupled simulation to capture the aero-acoustics from the flow around an airfoil. The geometry of the airfoil is placed in the innermost domain where the full compressible Navier–Stokes equations are solved. This is surrounded by a subdomain to capture the flow where viscous effects can be neglected and the inviscid Euler equations are solved. Finally, a far-field to capture the emitted sound waves is embedding these. In the far-field only the linearized Euler equations are solved, as no non-linearities are expected in this acoustic field anymore. Due to its linear nature a high spatial scheme-order can be employed in the far-field to reduce the memory that is required to cover the large domain. Thus, each subdomain utilizes different spatial discretizations and solves different equations. We present simulation results, showing the suitability of our coupling strategy for multi-scale problems and investigate the resulting load imbalances in the partitioned coupling. Furthermore, we highlight how we model the geometry in the high-order context and present how the performance can be improved using load balancing available in our simulation framework.

2   Geometry Modelling in High-Order

For the simulation of engineering applications in fluid dynamics an appropriate method is needed to model the geometry efficiently. When considering high-order methods for the discretization of the flow domain, the geometry has to be modeled in high-order as well to maintain the overall accuracy of the solution. Two methods are used, to introduce the geometry in the flow domain. The first method is based on adapting the mesh to the geometry and using special elements to account for its surface. The other method leaves the mesh untouched but introduces additional terms into the solved equations to address the geometry within the elements of the mesh.

Using the immersed boundary method without touching the mesh is especially useful for high-order discretizations with few elements or moving obstacles. However, it results in volumes that are not of interest for the solution but still need to be computed. We utilize this method to represent the geometry in our solver and minimize the effect of the additional computed volumes by minimizing the approximation where it is possible.

2.1   *Immersed Boundary Method*

The immersed boundary method is well suited for numerical simulations, to model not just simple but also arbitrary complex geometries. One major and outstanding advantage of this method is the possibility to employ elements independently from the geometry. This allows the free choice of element forms and enables the use of efficient numerical schemes. This enables us to utilize cubical elements that are very well suited for parallelization and efficient computation. Figure 1 presents the strategy used when considering the immersed boundary method to model the geometry on a Cartesian mesh. As demonstrated in the figure, the geometrical representation lives within the mesh, and no adaptation of the mesh towards the geometry is required. This is also useful if the geometry is to be moved during the simulation.

For our simulations the geometry is modeled via a Brinkman penalization [4], where it is represented by a porous material. This has the advantage, that we just need to include a few additional terms into our flow equations. The numerical discretization is based on the high-order modal discontinuous Galerkin method (DG), which allows high accuracy of the solution and faster convergence of the solution with less degrees of freedom (Dofs), when compared to a low order scheme. The

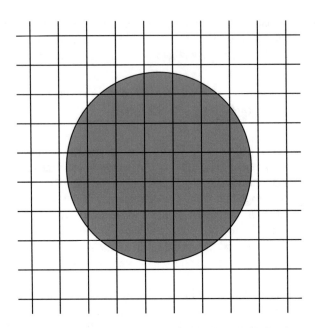

Fig. 1 Immersed boundary method: Geometry modeled on the computational cartesian mesh

physical state of the solution is represented in the model space for all conservative variables (mass, momentum and energy). Due to the low inherent dissipation and dispersion properties of this scheme, this method is well suited for aero-acoustic simulations. Furthermore, due to the integration of the geometry model into the flow equations with the same discretization as the solution, we can maintain the high accuracy of the overall solution and the high-order for our simulations. The equations below represent the conservation of mass (Eq. 1), momentum (Eq. 2) and energy (Eq. 3) for the compressible Navier–Stokes equations. The boxed part of the equations represent the additional terms we need to consider for the modelling of the geometry as a porous material. Where ϕ represents the porosity, χ the masking function, which is either 1 or 0, indicating whether geometry is present or not, U_{oj} the velocity of the obstacle, η the viscous permeability, η_T the thermal permeability and T_o the temperature of the geometry.

$$\frac{\partial \rho}{\partial t} = -\frac{\partial m_j}{\partial x_j} - \boxed{\left(\frac{1}{\phi} - 1\right) \chi \frac{\partial (u_j - U_{oj})\rho}{\partial x_j}} \tag{1}$$

$$\frac{\partial m_i}{\partial t} = -\frac{\partial}{\partial x_j}\left(m_i u_j\right) - \frac{\partial p}{\partial x_i} + \frac{\partial \tau_{ij}}{\partial x_j} - \boxed{\frac{\chi}{\eta}(u_i - U_{oi})} \tag{2}$$

$$\frac{\partial e}{\partial t} = -\frac{\partial}{\partial x_j}\left[(e + p)u_j\right] + \frac{\partial}{\partial x_i}(u_i \tau_{ij}) + \frac{\partial}{\partial x_j}\left(k\frac{\partial T}{\partial x_j}\right) - \boxed{\frac{\chi}{\eta_T}(T - T_o)} \tag{3}$$

A major drawback of the immersed boundary method besides its several outstanding advantages is the computation of the solution not just outside the geometry but also inside of it, where it is not of any interest. Therefore, additional computation is devoted to elements, which are out of interest for the overall solution. To tackle this shortcoming, we introduce a method to reduce the computational effort depending on the scheme order used and the number of elements, which are completely covered by the geometry.

2.2 Mode Reduction Method

Computation inside the geometry is not desired and should be avoided to minimize the computation effort. As mentioned in the above section, the modelling of the geometry using the immersed boundary method implies computation inside the geometry. Accordingly it would be better to cut out elements inside the geometry from the computational domain. However, this is not always easily possible and counters some of the benefits of this method, especially when considering moving obstacles. Alternatively, we can reduce the computational effort in completely immersed elements. We achieve this by a *mode reduction* in those elements. One of the most compute intensive operations when considering high-order DG method is the physical flux, since the computational effort of the physical flux computation increases with increasing scheme order and has to be done several times for each element. Therefore, we can decrease the computational cost, when taking this aspect into account, and fall back to the first mode (integral mean) of the polynomial representation of the state in elements that are completely inside the geometry. This means when an element is completely covered by the geometry, we just consider this first mode in the computation of the physical flux instead of computing all higher modes as well.

To decide whether we can make use of *mode reduction* for specific elements, we check the faces of the neighbouring elements and the element's own faces and check there, if the masking function χ has a value of 1 everywhere. In case the faces of the neighbouring elements and the element's own faces have the value 1 for χ we consider only the first mode in the physical flux computation for this specific element. With that we make sure, that we are not falling back in the accuracy of the solution in the boundary layer area and are consistent in the computation. Figure 2 gives a small overview how this new feature works in practice, when considering a small test case with 4 elements. As demonstrated in the figure, the blue element represents a fluid element, which is not covered by the geometry, while

Fig. 2 Reduction of mode for the physical flux computation, for elements covered by the geometry

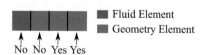

No No Yes Yes

■ Fluid Element
■ Geometry Element

the red marked elements are completely covered by the geometry located there. The *mode reduction* can only be applied for the last two elements, which have all faces inside the geometry as well as their neighbouring elements. While the third element is also completely covered by the geometry, but one of its faces has a neighbour which is not. Further the boundary layer inside the material needs to be properly resolved within this element. If we would reduce the spatial scheme order also in this element, the modelling of the wall would be badly affected.

3   Load Balancing

Coupled simulations not only allow the reduction of the overall computational effort for complex setups, but also enable the best-suited configuration for the occurring physics in each of the subdomains. However, a drawback of this approach is that load balancing gets more involved in this strategy. We can differentiate between intra-subdomain and inter-subdomain load imbalance. Load imbalance happening inside each subdomain (intra-subdomain), is due to the fact, that inside the subdomain those elements involved in the coupling at the coupling interface, for example, have more workload than other elements. Beside the computation of the equations, those elements have to maintain the communication to the coupling tool and exchange data (receive and provide) to update the state of the computational subdomain in each time step. Another example would be reduced costs by the *mode reduction* method described above in the presence of geometries. Since all subdomains have different sizes, solving different equations due to the occurring physics and considering different scheme orders, the workload varies from subdomain to subdomain, resulting in the need for inter-subdomain balancing. The imbalance introduced by the coupling approach and the load balancing of it is investigated in detail in [5].

The intra-partitioned workload stands even more out, due to additional load imbalance caused by the presence of geometries, which can e.g. move. As mentioned in Sect. 2, we consider an immersed boundary method, to model the geometry with the same accuracy as the underlying scheme. Due to the method used, we compute the solution also inside the geometry, where it is not of interest, hence resulting in additional computation. As mentioned in Sect. 2.2 by the means of the *mode reduction* feature we already counter the computational effort, but still need to account for the load imbalances induced by it. The next subsections are devoted to the differentiation of load imbalance (inter-subdomain and intra-subdomain) and the methodology we consider to overcome load imbalances for our simulations. The focus of this work is on intra-subdomain load imbalance and the workload resulting from the immersed boundary method to model the geometry by means of the *mode reduction* method.

3.1 Intra-Subdomain Load Imbalance

One important issue we face for our simulations is the workload distribution inside a subdomain. Figure 3 shows exemplarily intra-subdomain load imbalances and the balancing of it. In each scenario, there are 4 MPI processes depicted by the solid lines and 16 elements in total (four per MPI process) illustrated by the dotted lines. In this illustration each process works for the exact same time on each of its elements sequentially. At t_{sync} the processes have to wait for each other for synchronization. Figure 3a: ideally each process involved in the computation needs to have the same workload, so that all processes of one subdomain finish the computation of one time step at the same time. Each element in the domain has a defined workload, which in principle depends on the numerical scheme or its location. Figure 3b: coupled simulations are carried out via boundary conditions, resulting in surface coupling. Due to the coupling, elements at the coupling interface have additional load in terms of computation. At each time step, they have to provide the required coupling data at the required points in space. This might be costly for the chosen discretization. Further this data needs to be exchanged with the other subdomains as well. Here, we have 1 coupling element (light blue shaded box) on 1 MPI-process having additional coupling workload (blue box) and hence 3 MPI-process are idling at t_{sync} (gray shaded box) to wait for the coupling computation on the first process. Figure 3c: using a re-partitioning approach to distribute the elements within

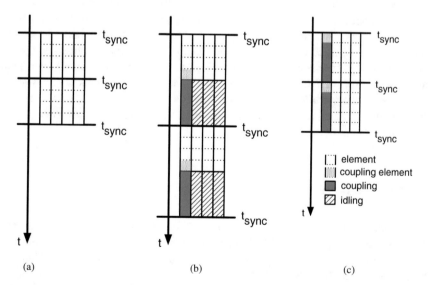

(a)   (b)   (c)

Fig. 3 Example of intra-partitioned load imbalances and load balancing within a subdomain. Solid lines depict 4 MPI-processes and the dotted lines illustrate 16 elements. (**a**) Ideal LB yielding same number of elements with all having same workload. (**b**) Same number of elements on each processes with one element involved in coupling. (**c**) Using re-balancing mechanics resulting in good load balancing with one element involved in coupling

a subdomain according to their workload. Here 1 expensive coupling element on 1 MPI-process and 5 elements on the 3 other MPI-processes.

The consideration of reduced workloads due to the described *mode reduction* inside geometries results in similar imbalances within subdomains, contributing to the heterogeneity of the workload distribution. Therefore, we need to consider the load imbalance caused by the decomposition of the domain and the additional imbalance due to the *mode reduction* feature accordingly, to improve the computational efficiency of the partitioned coupling setup. The concepts of load imbalances explained for coupling in Fig. 3 can be transferred to any kind of load imbalances. It has to be mentioned, that whenever intra-subdomain load imbalance is present, the computation of the coupled scenario also influences the inter-subdomain load imbalance, as waiting time for other subdomains (inter-subdomain load imbalance) is inevitable.

3.2 *Inter-Subdomain Load Imbalance*

Besides intra-subdomain load imbalance, waiting times can also arise between subdomains. Different equations, spatial domains and numerical discretizations lead to different workloads in the individual subdomains. For example solving a nonlinear Euler flow with around 1000 fine elements and lower order has a different workload than a linearized acoustic field with 500 coarse elements and a higher order. Each subdomain receives before execution a predefined number of processes that does not vary over runtime. For a good load balancing, the number of processes should be according to the total workload (computation of the equation and spatial order, coupling, geometry computation) of the subdomain. Inter-subdomain load imbalances are not the focus of this work. However, it influences the overall performance and needs to be taken into account as well. More information regarding Inter-subdomain load imbalance due to coupled simulations and how to deal with that can be found in [5].

4 Methodology

This work is done in our in-house simulation framework APES [6, 7], where the integrated coupling approach APESmate [8] is included. To address load imbalances in our simulations, we make us of the integrated space-filling curve based balancing algorithm SPartA [9]. This feature is found in the common mesh library of APES called TreElM [10]. SPartA is based on the space-filing curve and individual weights which provides information regarding the actual load per element. This algorithm is well suited for static meshes, in terms of re-partitioning. The weights are determined by time measurements during runtime of the solver. Timers are set before the general computation, which includes among others the physical flux computation and the projection onto the test function, which are the most expensive operations. Additionally timers are included for elements involved in the coupling, which have additional load to manage. The weights written out after a successful run are independent from the number of processes used for the simulation, since they provide information regarding the actual work each element has. Since timers are used for the measurement, the workload has the time dimension and is given in seconds [5]. Through the weights we ensure that the re-partitioning among the processes is according to the actual work load, which includes all variations from element to element.

5 Results

This section is devoted to simulation results as well as results showing the performance of the coupled simulation when compared to the monolithic approach. We also show how we can save computing time, when using the integrated load balancing approach for a 3-field coupled simulation. The coupled scenario is consisting of an innermost subdomain, where an airfoil (S834) profile is located, which is modeled as a porous material (see Sect. 2). Here we consider the full compressible Navier–Stokes equations, using a fine mesh and a low scheme order to resolve small scales and consider the viscosity. Away from all viscous effects, we simplify our equations and solve the compressible inviscid Euler equations (middle subdomain), where a coarser mesh and a higher scheme order is used. Further away from all nonlinearities we have the third subdomain, where we are only interested in the propagation of acoustic waves. Therefore, we solve the linearized Euler equations were an even coarser mesh and an even higher scheme order is used. For all simulations we use the high-order Discontinuous Galerkin (DG) solver Ateles and individual subdomains via APESmate. With this test case, we utilize the properties of DG of low dissipation and dispersion error, which is of importance for the transportation of information over larger distances. This strategy allows us to reduce the computational effort and consider expensive equations where it is needed and simplify the equations as soon as the physics allows.

5.1 Configuration of the Simulation

The simulation configuration for the 3-field coupling is shown in Table 1. From the table it is clear that the smallest volume is covered by the innermost domain, where we solve the expensive Navier–Stokes equations. At the same time this domain contains the most elements. This is necessary to capture small scales, but also to resolve the boundary layer at the geometry interface. Furthermore, the outermost domain, where the linearized Euler equations are solved has the least elements, while using a spatial scheme order of 9, which is the highest for this test case. Due

Table 1 Setup for 3-field coupled simulation

	Navier–Stokes domain	Euler domain	Linearized Euler domain
	Inner domain	Middle domain	Outer domain
Domain length [x, y, z]	[6, 4, 4]	[12, 10, 4]	[12, 5, 4]
Number of elements	450,420	24,576	3,840
Scheme order	4	6	9

to the high-order scheme we can make sure to have low dissipation and dispersion over the large domain. The domain size is provided in Table 1. All lengths are normalized by the cord length of the airfoil. In the Navier–Stokes domain a jet-inlet is prescribed, which injects a direct stream on the airfoil geometry. The jet at the inflow is located exactly in the middle and has a radius of 0.5 unit length. The kinematic viscosity μ has a value of 10^{-7}. The velocity at the jet-inflow is ramped linearly over the time and has a value of **velocity** $=[0.1 \cdot \sqrt{(\gamma \cdot p/\rho)}, 0.0, 0.0]$, with γ equal to 1.4, pressure p being 101,325 and density $\rho = 1.0$. The values for pressure and density are also set as initial conditions for the middle domain and as background values for the outer domain, while the perturbation is defined to be 0.0 initially for all state variables (pressure, density and velocity) for the outermost subdomain.

Our purpose is, to show that we can obtain a much faster computation by decomposing the simulation domain and using load balancing to distribute the workload among available resources accordingly. Since this test case is too big to be also computed monolithically for comparison, we reduce the domain size in a way, that a comparison between the performance when computing the entire simulation domain with the same equations, scheme order and spatial discretization (monolithic) with the coupled scenario, is possible. Therefore, we consider Table 2 for the performance comparison. For the monolithic approach we use the same domain size as for the coupled simulation, while solving the entire domain with the Navier–Stokes equation, a scheme order of 4 and the same element size as in the innermost subdomain of the coupled test case, hence resulting in 1,216,748 elements and 389,359,360 degrees of freedom (nDof), which is more than what is used for the coupled scenario, which is discretized with 43,291,520 degrees of freedom in total. A more detailed look into the number of degrees of freedom reveals that the coupled test case has 9 times less degrees of freedom than the monolithic test case. This

Table 2 Small setup for 3-field coupled simulation

	Navier–Stokes domain	Euler domain	Linearized Euler domain
	Inner domain	Middle domain	Outer domain
Domain length [x, y, z]	[4, 2, 2]	[12, 6, 2]	[12, 3, 2]
Number of elements	94,516	8192	1152
Scheme order	4	6	9
nDof	30,245,120	8,847,360	4,199,040

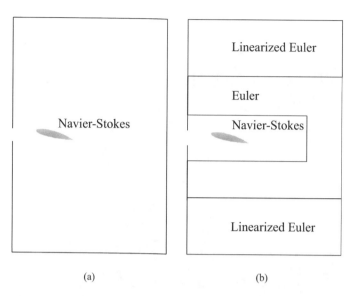

Fig. 4 (**a**) Monolithic test case and (**b**) coupled test case

is of importance, when considering the evolution of new supercomputing systems towards decreasing memory per node and thus the increased importance of memory consumption.

In Fig. 4 the test case is shown, while computing the entire domain in (a) monolithically and in (b) decomposed. As mentioned previously a drawback of coupled simulations is the fact, that load imbalances are introduced inside the domain as well as from one domain to another one. In order to reduce computational overhead, load balancing is indispensable and has to be addressed accordingly. The first level of load imbalance that needs to be addressed is the different load each element has to take care of (intra-subdomain load imbalance).

5.2 Weight Distribution

For load balancing purposes the load each element has to deal with depends on its location as well as if e.g. source terms are available in them. In the coupled scenario additional workload is also introduced at the coupling interfaces due to the necessary interpolations. The weights are obtained by element timers that are placed around compute intensive routines e.g. the physical flux computation or the evaluation of the geometry and make use of MPI_Wtime. Figure 5 shows the weights for each element of a small coupled test case exemplarily. The test case consists of an inner domain solving the Navier–Stokes equations, where a cylinder is located in the center of the domain with a radius of 1.0. The outer domain is solving the inviscid

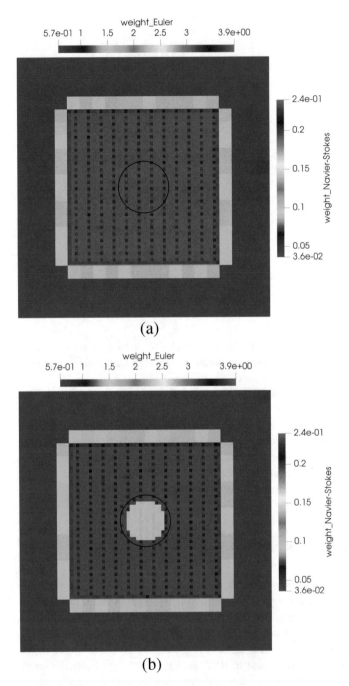

Fig. 5 Results of the workload for a small coupled scenario. Figures showing the inner domain (Navier–Stokes) and the outer domain (Euler) in (**a**) weights, showing the dominant coupling elements in the outer domain and (**b**) reduction of the weights for elements inside the geometry when using *mode reduction* in the inner domain. The white frame depicts the coupling interface between the two domains and the black line the cylinder location inside the inner domain

Euler equations. In the inner domain the edge length of the elements is a fourth of the edge length of the elements in the outer domain. As can be clearly seen, the inner domain has mostly evenly distributed weights throughout the domain, while the outer domain has very compute intensive elements at the coupling interface (indicated by the white line). This is due to the information it has to provide to the inner domain. Since the inner domain (Navier–Stokes) requires not only the state variables, but also their gradients, those elements become even more computational intensive. While the inner domain just needs to provide the state values to the outer domain, and hence the additional computation is nearly invisible in the weights. In the inner domain some regular pattern can be recognized, this regular pattern is likely due to memory access patterns. The effect of the *mode reduction* discussed in Sect. 2.2 can be seen in Fig. 5b, here the workload is clearly reduced within the cylinder indicated by the black line. Figure 5a shows the weight distribution if no *mode reduction* is used, illustrating a nearly even distribution of weights but wasted computational effort within the geometry region.

Note that there are two different scales used in Fig. 5 to account for the different element sizes. There are 16 elements in the inner domain covering the same area as a single element in the outer domain. Thus, there is a factor of 16 used in the scales to better illustrate the computational intensity per covered simulation volume. With the knowledge obtained to this point we can conclude, that the coupling elements can be computationally intensive. Furthermore, we can reduce the workload of the elements inside the geometry by the newly implemented feature to address the unnecessary expensive computation of elements located inside the geometry by reducing the utilized polynomial modes in their computation to 1.

5.3 Performance Measurements

To investigate the performance of the mentioned test case in Sect. 5.1 and find out how much performance improvement we can obtain by considering load balancing, we compare the coupled simulation with and without load balancing to the monolithic approach, where we consider both scenarios of load balancing as well. In all cases we make use of the *mode reduction* feature, to neglect the computation inside the airfoil. Furthermore, we consider the element timers to write out weights during runtime, which then can be used to balance the load using the SpartA algorithm and computing the number of cores each subdomain needs for the computation in order to avoid idling overheads. The load balancing is done considering intra- and inter-subdomain workload. The investigations were done on the Noctua system, a Cray installation in Paderborn, Germany. Figure 6 illustrates the strong scaling measurements when considering up to 2560 processes for the investigation, with each node having 40 processes for the execution of the simulation. It is noticeable how much performance benefits we can obtain by decomposing the domain and the different treatment. Even though the coupling elements are very expensive when compared to the elements inside the domain,

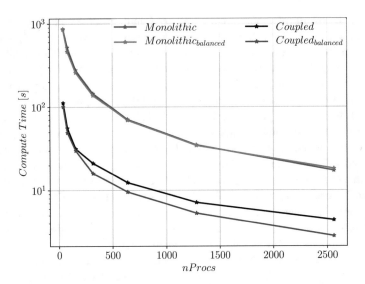

Fig. 6 Comparison of the strong scaling measurement for the 3-field coupled simulation-with and without load balancing-and the monolithic test case

still the coupled scenario provides outstanding strong scaling results. We can also observe, that since for the monolithic approach almost all elements have the same workload, we can not gain much benefit from the load balancing used here, since the geometry covers a small amount of elements, which are responsible for additional load imbalance, when using *mode reduction*. While for the coupled scenario the performance improvement using the introduced load balancing method is significant and allows further performance improvement for the already very efficient coupled simulation. For both coupled scenarios (with and without load balancing) inter-subdomain load balancing is used, while for the not balanced scenario the intra-subdomain load balancing was neglected. For inter-subdomain load balancing we started with 36, 3 and 1 processes, where the innermost domain ran on 36 processes, the middle domain on 3 and the outermost domain on 1 process. This ratio was kept for strong scaling measurement. Hence for the run with 80 processes we used 72, 6 and 2 processes for each subdomain respectively.

5.4 Simulation Results of the Coupled Scenario

The simulation results from the coupled scenario are shown in Fig. 7. At the inlet a jet-inflow is prescribed, which provides a stream around the airfoil located in the most inner subdomain. From the figure it is clearly visible, that vortices in the innermost subdomain are well resolved and travel from the inner to the middle subdomain, hence passing the coupling interface without changing their properties

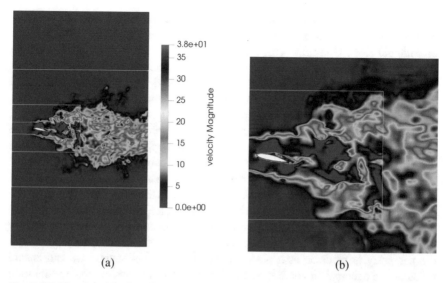

(a) (b)

Fig. 7 Results of the coupled simulation in (**a**) the velocity magnitude over the domain and (**b**) a zoom-in at the coupling interface between the most inner domain (Navier–Stokes) and the middle domain (Euler). The white frames mark the coupling interfaces between each subdomain

and still maintaining the physics correctly. The white lines represent the coupling interface, where the data-exchange between the coupling domains occurs. Even though each subdomain solves a different set of equations and the mesh resolution as well as the scheme order are different (see Table 1), there is no shifting of physical attributes or the traveling of wrong information from one domain to the other observable. Further, vortices are well resolved in the inner domain, where it is needed. It has to be mentioned, that the flow is still not completely evolved and has to run further to see also the acoustics propagation into the outermost subdomain. Due to the right choice of equations and the overall simulation setup, all scales can be represented accurately. In the zoom-in picture we can not only see how well the physics are preserved but also the change in the mesh resolution, which can be addressed during the post-processing procedure.

6 Conclusion

Complex simulations involving multi-scale and multi-physics are still challenging when considering engineering applications. Computing these setups in a monolithic manner is not feasible from the computational perspective. In order to tackle this problem, we make use of partitioned coupling, where we split the domain according to the physical requirements and solve each of the domains with a tailored configuration. Subdomains are weakly connected to each other via the boundaries, where the data-exchange between them takes place through a coupling tool. With this strategy we can not only obtain accurate simulation results, but also enable runs in terms of compute time. One drawback we face when considering partitioned coupling is the increased complexity in the arising load imbalances. We can distinguish here between two levels of load imbalance the intra- and inter-subdomain load imbalance. The intra-subdomain load imbalance results from the different loads each element inside a subdomain might have. For coupled simulations in particular, this is caused due to the coupling elements at the coupling interface, which have to maintain the communication between the coupling approach and are responsible for providing and receiving data from and to the coupling tool. This drawback is addressed in this paper. We described how we deal with both levels of load imbalance and presented how we make use of element timers in the solver to obtain weights per elements, which can then be used to balance the workload appropriately among available processes. Furthermore, we presented how we model the geometry for engineering applications by means of an immersed boundary method, where the geometry is modeled as a porous material. With this method we can ensure efficient and accurate simulation results, but need also to consider that the solution is also computed inside the geometry, where it is not needed. By reducing the spatial scheme order in those elements, this disadvantage can be minimized, but again load imbalances are introduced within the domain. Furthermore, we presented outstanding performance results for the coupled scenario by means of a 3-field coupled simulation with an airfoil located inside the innermost domain. The strong scaling measurements show how important the load balancing is for the scalability of the coupled scenario. We also showed that we can simulate the physics according to its need by the means of a 3-field coupled problem, where we decomposed the entire simulation domain in three subdomains and solved them with the best-suited numerical configuration.

The future work is devoted to the investigation of load imbalance, when considering not just stationary geometries but also moving ones. For moving geometries the masking function of the penalization needs to be evaluated in every time step, resulting in increased costs in domains with moving geometries. In addition we need to further study how large the benefit from using the *mode reduction* feature for moving geometries is, as well.

Acknowledgements This work was financially supported by the priority program 1648 - Software for Exascale Computing 214 (www.sppexa.de) of the German Research Foundation. The performance measurements were performed on Noctua system at Paderborn Center for Parallel Computing (PC2). Simulation results were obtained on the Supermuc supercomputer at Leibniz Rechenzentrum (LRZ) der Bayerischen Akademie der Wissenschaften. The authors wish to thank for the computing time and the technical support.

References

1. Pour, N.E., Krupp, V., Klimach, H., Roller, S.: Coupled simulation with two coupling approaches on parallel systems. In: Resch, M.M., Bez, W., Focht, E., Gienger, M., Kobayashi, H. (eds.) Sustained Simulation Performance 2017, pp. 151–164. Springer International Publishing, Cham (2017)
2. Pour, N.E., Roller, S.: Error investigation for coupled simulations using discontinuous Galerkin method for discretisation. In: Proceedings of ECCM VI / ECFD VII, Glasgow (2018)
3. Schwartzkopff, T.: Finite-Volumen Verfahren hoher Ordnung und heterogene Gebietszerlegung für die numerische Aeroakustik. PhD Thesis, Universität Stuttgart, Institut für Aerodynamik und Gasdynamik (2005)
4. Liu, Q., Vasilyev, O.V.: A brinkman penalization method for compressible flows in complex geometries. J. Comput. Phys. **227**(2), 946–966 (2007)
5. Krupp, V.: Efficient coupling of fluid and acoustic interactions on massive parallel systems. PhD Thesis, Universität Siegen, to be published in 2020
6. Zudrop, J., Klimach, H., Hasert, M., Masilamani, K., Roller, S.: A fully distributed CFD framework for massively parallel systems. In: Cray User Group 2012, Stuttgart (2012)
7. Roller, S., Bernsdorf, J., Klimach, H., Hasert, M., Harlacher, D., Cakircali, M., Zimny, S., Masilamani, K., Didinger, L., Zudrop, J.: An adaptable simulation framework based on a linearized octree. In: Resch, M., Wang, X., Bez, W., Focht, E., Kobayashi, H., Roller, S. (eds.) High Performance Computing on Vector Systems 2011, pp. 93–105. Springer, Berlin (2012)
8. Krupp, V., Masilamani, K., Klimach, H., Roller, S.: Efficient coupling of fluid and acoustic interaction on massive parallel systems. In: Sustained Simulation Performance 2016, pp. 61–81. Springer, Cham (2016)
9. Harlacher, D.F., Klimach, H., Roller, S., Siebert, C., Wolf, F.: Dynamic load balancing for unstructured meshes on space-filling curves. In: 2012 IEEE 26th International Parallel and Distributed Processing Symposium Workshops PhD Forum (IPDPSW), pp. 1661–1669 (2012)
10. Klimach, H.G., Hasert, M., Zudrop, J., Roller, S.P.: Distributed octree mesh infrastructure for flow simulations. In: Eberhardsteiner, J. (ed.) ECCOMAS 2012 - European Congress on Computational Methods in Applied Sciences and Engineering, e-Book Full Papers (2012)

Part V
Applications and Numerical Methods for Large-Scale Systems

Large Scale Agent Based Social Simulations with High Resolution Raster Inputs in Distributed HPC Environments

Sergiy Gogolenko

Abstract Agent-based modelling and simulation (ABMS) is an essential tool which allows to explore the role of social phenomena via computer simulation. Large scale social simulations of HPCs—also known as parallel and/or distributed agent-based simulation (PDABS)—play a key role in the emerging field of computational global systems sciences (GSS). Agent-based models (ABMs) in GSS are characterized by highly non-uniform spatial distribution of agents and importance of long distance social interactions. Over the last two decades, researchers proposed a number of approaches to effectively address these traits of ABMs. Such approaches are driven by data available for scientists. In many GSS applications, the data partially come in raster formats. Yet, case of raster inputs for GSS applications is barely studied in literature and not supported sufficiently in the state-of-the-art ABMS frameworks. In this paper, we propose a graph-based approach to represent ABMs with raster inputs on HPCs. This approach naturally leads to a space-relationships-based work partitioning strategy which allows to improve performance of PDABS.

1 Introduction

Large scale social simulations play a key role in the emerging field of computational global systems sciences (GSS) which deals with providing "scientific evidence to support policy-making, public action, and civic society" [1]. to provide scientific evidence to support policy-making, public action and civic society to collectively engage in societal action. "The behaviour of many social systems requires that they be modelled at the level of individual people" [2], which is usually achieved by agent-based modelling and simulation (ABMS). On the one hand, since GSS applications analyze society on global or country level, individual-based view on

S. Gogolenko (✉)
High Performance Computing Center Stuttgart (HLRS), Stuttgart, Germany
e-mail: gogolenko@hlrs.de

© Springer Nature Switzerland AG 2020
M. M. Resch et al. (eds.), *Sustained Simulation Performance 2018 and 2019*,
https://doi.org/10.1007/978-3-030-39181-2_16

the global systems naturally leads to computationally expensive large scale ABMS runs. On the other hand, the modeler obtains flexibility in addressing heterogeneity in agents, non-linearity in their responses, and other complex model assumptions. This flexibility enables ABMS to capture emergent social phenomena overlooked by macro- and meso-scale models [3], as well as to outperform conventional machine learning techniques in some cases [4].

At the same time, agent-based models (ABMs) for GSS have many peculiarities. In [5], authors list three major traits of ABMs in GSS, which differ them from ABMs encountered in other scientific domains such as computational biology or ecology. These are heterogeneity of agents, highly non-uniform spatial distribution of agents in the environment, and important role of long distance (social) interactions. Heterogeneity of agents reflects diversity of actors involved into GSS models. Non-uniform spatial distribution of agents is caused by urbanization processes, obstacles imposed by nature, etc. ABMs in GSS often encompass two types of communications—short distance communications representing interactions due to spatial proximity of agents, and long distance interactions standing for social relationships. In many cases, the latter dominate over the former.

Over the last two decades, researchers proposed a number of models and data structures to effectively address traits of ABMs in GSS on HPC clusters. These models and data structures are driven by data available for scientists. The data about environment may come in vector or raster format. ABMs with inputs in vector formats cover the majority of GSS use cases. Examples of recent developments focused on inputs in vector formats include, but not limited to hierarchical (tree-like) models [6, 7], directed probabilistic social networks [8], social contact networks [9], urban geo-social networks [10]. At the same time, in many applications the global system scientists use the data about ABM environment available in raster formats from popular open data sources like NASA's Socioeconomic Data and Applications Center [11] and others. In this paper we propose an HPC compliant model and corresponding data structure for this situation.

The rest of the paper is organized as follows. Section 2 briefly reviews state-of-the-art HPC compliant ABMS software for global system scientists, as well as approaches to model spatial environment and distribute workload implemented in these tools. Section 3 presents graph-based HPC compliant model and corresponding data structure for ABMs with raster inputs and strong long distance interactions between agents. Section 4 illustrates performance of our solution on the toy benchmark implementing Axelrod's model of dissemination of culture. Finally, Sect. 5 discusses conclusions and direction for further work.

2 Previous Work

During the last decade, a number of HPC compliant ABMS codes were developed [12–14]. These developments vary from domain specific tools to general purpose ABMS frameworks.

Although domain specific tools usually target individual use cases, they address traits of particular GSS applications very effectively. Moreover, many ideas implemented in such tools have generic nature and can be applied to a broader number of use cases. The remarkable examples of the domain specific tools are FluTE[7], EpiFast[8], EpiSimdemics[9]. FluTE[7] represents the model of iterations in society as a multi-level tree. The root of the tree corresponds to the whole society, while the lower levels of the tree represent elements of society with finer granularity until reach the level of individual households as leaves. The probability of interactions between agents is dictated by the distance to the closest common parent. EpiSimdemics[9] implements a so-called social contact network (SCN) model. In SCN, society is represented by an affiliation (bipartite) graph with agents on one side and loci of their interactions (environment) on the other. This affiliation graph is accompanied with a schedule of interactions. Both SCN and tree-like models take into account only short range interactions between agents. Urban Geo-Social Network Model (UGSN) proposed in [10] addresses this limitation by further development of the SCN idea. It represents the society by SCN and additional multilayer network[15] of direct social connections between agents for modelling long distance communications.

Noticeable examples of the HPC compliant general purpose ABMS frameworks are RepastHPC[16], D-MASON[17], Flame-GPU[18], and Pandora[19]. Despite the wide choice, these frameworks fail to address all common traits of GSS applications effectively. Being implemented in Java, D-MASON has limited potential for use and porting on state-of-the-art large-scale HPC clusters. FlameHPC and Pandora lack proper support for simulation of social connections between agents. Although RepastHPC has formally all components required to build ABMs in GSS, it demands significant HPC expertise and advanced programming skills from the modeler (due to intricate and verbose API). Moreover, latest version of RepastHPC still ignores recent advances in high performance data structures such as new techniques for handling evolving graphs, modern fast implementations of hash tables, etc.

The common bottleneck for the majority of general-purpose ABMS frameworks is a naïve approach to model spatial environment and distribute workload. Many popular frameworks—including Flame-GPU, RepastHPC, and Pandora—model environment topology by cartesian grids. In 2D case, environment attributes are represented by dense matrices of the same size. Indices of the matrices correspond to the spatial coordinates and define locations of the grid vertices. During the distributed simulations, cartesian grid is split evenly between processes (Fig. 1a). This approach is often referred as uniform partitioning[20]. Since amount of computational work in agent based simulation step is proportional to the number of agents, uniform partitioning results in a significant load imbalance if agents are distributed very non-uniformly in space. As a result, this approach allows to reach reasonably good performance for many classical ABMS applications, but gives poor performance in situations with non-uniform spatial distribution of agents which is a case of GSS applications where agents are highly concentrated in the urban areas and sparsely distributed outside the settlements. D-MASON tackles this

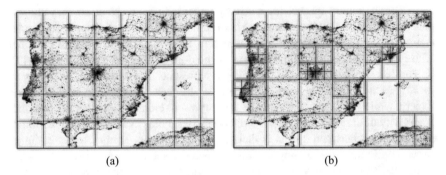

Fig. 1 Approaches to partition the rasters. (**a**) Uniform partitioning. (**b**) Non-uniform (tree codes based) partitioning

limitation by introducing a space-based non-uniform work partitioning approach based on tree codes (multi-scale meshes) [20]. This approach extends idea of the quad-tree Barnes-Hut algorithm, widely used in n-body simulations, to the agent-based models [21]. In particular, in [20], authors propose to use a so called bounded pseudo quad-tree (Fig. 1b). Even though the space-based work partitioning approach significantly reduces load imbalance, it does not take into account the situation when long distance communications play an important role in social simulations. As discussed in Sect. 1, the latter refers to the vast majority of GSS applications.

3   Graph-Based Model for Sparse Raster Inputs

Concept

Figure 2 illustrates typical organization of inputs for ABMs where data about environment comes in raster format. The information from the rasters can be combined into sparse spatial graph. Vertices of the spatial graph correspond to the non-empty pixels of the rasters and represent sites populated by agents. Each site is attributed with a tuple of pixel values in the corresponding position for all raster. Edges of the spatial graph stand for spatial proximity between sites and serve to model short distance communications. Agents are linked into a multilayer network of direct social connections. In addition, each agent is assigned to the site corresponding to its spatial location. This results into internal representation where agents linked into multilayer network of social connections $G_A = (V_A, E_A)$ are mapped on the sites linked into spatial graph $G_S = (V_S, E_S)$ (see right side of Fig. 2).

In order to keep workload balanced, the spatial graph should be distributed between processes taking into account the number of agents in sites, as well as short and long distance communications between agents. It can be achieved if we map multilayer network of social connections on the spatial graph to obtain

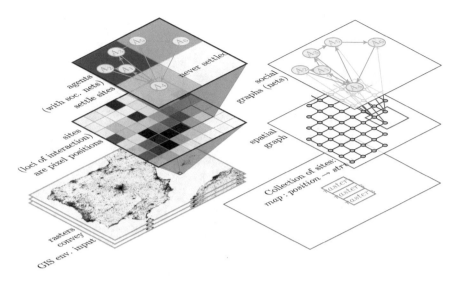

Fig. 2 Internal representation of ABMs with raster inputs

a computational graph $G_c = (V_c, E_c)$ with vertices $v \in V_c$ corresponding to sites and edges $e \in E_c$ corresponding to short and long distance communications. In this graph, weight of the vertex w_v equals the number of agents located at the corresponding site, while weight of the edge w_e between vertices equals the total number of agents in both vertices if sites are spatialty connected or the number of social links between agents assigned to these vertices if sites are distant. Optimal partitioning of the computational graph gives balanced distribution of agents between processors.

However, if rasters have high resolution, this approach cannot be used directly in distributed HPC environments since the number of sites becomes too big to address them effectively and to perform a balanced partitioning of the computational graph. This obstacle can be overcome by grouping sites into the chunks and partitioning the computational graph built upon the chunks of sites instead of individual sites. Nevertheless, even spatial partitioning of the sites into chunks might significantly disbalance the number of agents assigned to chunks. The better way is to build chunks upon tree codes—quad-trees or bounded pseudo quad-trees— using approaches discussed in [20]. The latter leads to a graph-based model and corresponding data structure illustrated in Fig. 3a. Note that this model uses the data structure similar to the data structure behind combination of USGN model with hierarchical tree-like model (see Fig. 3b). According to the taxonomy of PDABS work partitioning strategies proposed in [20], the corresponding work partitioning approach belongs to the class of space-relationships-based strategies.

Software for Implementation
Neither of the data structures depicted in Fig. 3 can be implemented in existing HPC compliant ABMS frameworks without dramatic changes in their cores. In

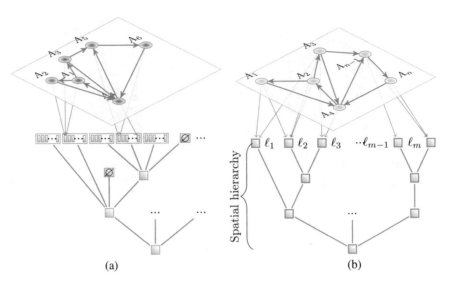

Fig. 3 Comparison of data structures of fine grained graph-based models designed for ABMs with sparse raster inputs and with vector inputs. (**a**) Graph-based with sparse raster inputs. (**b**) UGSN + hierarchical

Table 1 Support of short distance and long distance interactions between agents in Pandora and RepastHPC

	Pandora	RepastHPC
Social relationships representation		
Type of model	None	Directed multilayer network
Evolving social graphs		+
Edge weights		+
Multiplicity		+
Implementation		Native (ptrs)
Spatial component representation		
Format	`std::map` of dense matrices	Grid projector with vector of dense matrices (value layers)
Boundary conditions	Von Neumann	Von Neumann, Moore
GIS support	Rasters with GDAL	None

order to support this claim, Table 1 compares the most advanced HPC compliant frameworks written in C++—Pandora and RepastHPC—with respect to coverage of features necessary to implement the approach discussed above. This comparison shows that Pandora does not support multilayer network of social connections, whereas RepastHPC has insufficient number of instruments to implement spatial graphs.

In order to implement the approach without ABMS frameworks, one needs a graph partitioning tool and a general purpose graph library, which provides functionality sufficient to model social relationships.

The most appropriate general purpose graph libraries are Snap[22], PBGL, and GraphLab/PowerGraph[23]. PBGL (Parallel Boost Graph Library) is a rather lightweight package which supports most of the features required to model multilayer network of social connections and spatial graph. Even though interface of PBGL is designed for the users with advanced CS-skills, VTK library provides easy-to-use wraps over native PBGL interfaces. PowerGraph is an advanced distributed framework which implements graph-based gather-apply-scatter (GAS) programming model[23]. On the one hand, concept of the graph-based ABM simulation maps perfectly on the GAS programming model. In particular, "apply" phase allows to specify behaviour of the agents, and "gather" phase allows to collect suitable information from neighbours. On the other hand, PowerGraph does not support dynamic changes in the graph structure (vertices removal, etc). The latter strongly limits potential use of PowerGraph for ABMS. Both PBGL and PowerGraph assume that all vertices must have the same attributes. Nevertheless, the effect of versatility in vertex attributes can be achieved with variant types.

The incomplete list of remarkable graph partitioning packages developed over the last decades includes PT-Scotch, ParMETIS, PaGrid, Chaco, JOSTLE, MiniMax, ParaPART, DRUM, etc. But two of them—METIS and Scotch—gained much more popularity than others and are often referred as load balancing tools of choice in sophisticated time-consuming parallel numerical simulations. While both packages fit well to the needs of graph-based approach, ParMETIS is preferable since it allows to repartition distributed graph dynamically.

4 Benchmark for a Proof-of-Concept Implementation

In order to assess performance of our solution, we prepared a toy benchmark that implements Axelrod's model of dissemination of culture. This model was proposed in 1996 by R. Axelrod[24] and immediately gained broad popularity among social scientists. Nowadays, it is considered as one of the most well studied ABMs—both theoretically and empirically[25, 26],—which motivated us to choose Axelrod's model for benchmarking.

The model defines agents and rules for their interactions as follows. Agents model individuals in culture dissemination process. Each agent is endowed with F integer attributes called cultural traits, which are meant to model different beliefs, opinions, and other properties of agents. The model allows only a limited number of values for each cultural trait $f_i = (0, 1, \ldots, q_i - 1)$. In the dynamic step, each agent randomly selects one neighbour and the agent interacts with the neighbour with some probability proportional to the overlaps between the agent-neighbour pairs (the overlap is computed as a number of equal features). The interaction consists in assigning to one of the agent's trait the value of its neighbour trait. In other words, these rules make interacting agents more similar, but the interaction happens more often if agents already share many traits and it never happens if agents have no trait in common. This suggests that Axelrod's interaction rules allow

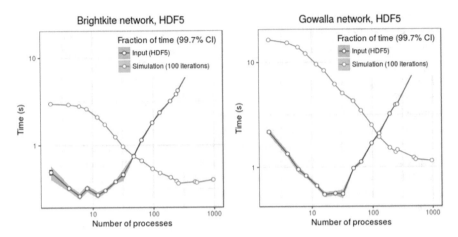

Fig. 4 Scalability of Axelrod's model implemented with the Amos framework on Hazelhen cluster

to model two cultural mechanisms—social influence and homophily. In order to fit Axelrod's model to the graph-based approach discussed in Sect. 3, we slightly modified Axelrod's notion of neighbours. In our implementation, we consider as neighbours all agents located at the same spatial site, as well as the agents that have direct links in the social graphs.

The benchmark was performed on the Hazelhen cluster at HLRS. Hazelhen is composed of CRAY XC40 nodes and has peak performance 7.4 Pflops. The cluster includes 41 Cray cascade cabinets with in total 7712 dual socket compute nodes. Each node is equipped with 2 12-core Intel Haswell E5-2680v3 CPUs and 128 GB of DDR4 RAM. The type of interconnect is Cray Aries. It uses Lustre storage and operates with the Cray Linux Environment. In our implementation, we used Snap 3 library as a back-end general-purpose graph library, and METIS as a graph partitioning tool. We compiled all components with GCC 6.4. Input files were pre-processed into CSV-format.

In our benchmarks, we used two networks from [22]—Brighkight and Gowala—for representing long-range interactions. The number of agents was artificially adjusted to the number of vertices in the networks. In order to create sites for their allocation, we used a 240×290 pixel raster with a gridded population density heat map of the Faroe islands from NASA's SEDAC. Agents were initialized with three cultural features each taking random values between 0 and 9.

Figure 4 summarizes results of the benchmark. Its subplots contain line chart with confidence intervals for measured elapsed times of data input and 100 simulation iterations of Axelrod's model. In these plots, we compare performance of the simulation iterations against embarrassingly parallel data input on both social networks. By embarrassingly parallel CSV input, we mean a naïve embarrassingly parallel implementation of the CSV reader which assumes that the user split the input data and prepared CSV files for each MPI process separately. For both networks, the scalability of the simulation iterations is better than the scalability of embarrassingly parallel input.

5 Conclusions and Further Work

The case of raster inputs for GSS applications is barely studied in literature and not supported sufficiently in the state-of-the-art ABMS frameworks. In this paper, we proposed a new graph-based model and corresponding data structure for efficient implementation of ABMs with raster inputs on HPC clusters. This model combines ideas of UGSN model, hierarchical models, and tree codes. State-of-the-art ABMS frameworks do not provide sufficient features to implement such model out-of-the-box. Nevertheless, we have shown that the model can be efficiently implemented with the general-purpose graph libraries and graph partitioning tools.

In the future, we plan to include support of this model in one of the ABMS frameworks, as well as to assess performance of our model on the large scale real world use cases.

References

1. EC: Global systems science. Internet (2017). Accessed 24 Jun 2019
2. Dum, R., Johnson, J.: Global systems science and policy. In: Non-equilibrium Social Science and Policy. Understanding Complex Systems, pp. 209–225. Springer International Publishing, Cham (2017)
3. Farmer, J.D., Foley, D.: The economy needs agent-based modelling. Nature **460**(7256), 685–686 (2009)
4. Suleimenova, D., Bell, D., Groen, D.: A generalized simulation development approach for predicting refugee destinations. Sci. Rep. **7**(1), 13377 (2017)
5. Paolotti, D., Tizzoni, M., Edwards, M., Fürst, S., Geiges, A., Ireland, A., Schütze, F., Gesine, S.: D4.1-first report on pilot requirements. Deliverable 4.1, CoeGSS—Centre of Excellence for Global Systems Science (2016)
6. Germann, T.C., Kadau, K., Longini, I.M., Macken, C.A.: Mitigation strategies for pandemic influenza in the United States. Proc. Natl Acad. Sci. **103**(15), 5935–5940 (2006)
7. Chao, D.L., Halloran, M.E., Obenchain, V.J., Longini, I.M.: Flute, a publicly available stochastic influenza epidemic simulation model. PLoS Comput. Biol. **6**(1), e1000656 (2010)
8. Bisset, K.R., Chen, J., Feng, X., Anil Kumar, V.S., Marathe, M.V.: EpiFast. In: Proceedings of the 23rd International Conference on Supercomputing - ICS '09 (2009)
9. Barrett, C.L., Bisset, K.R., Eubank, S.G., Feng, X., Marathe, M.V.: Episimdemics: an efficient algorithm for simulating the spread of infectious disease over large realistic social networks. In: 2008 SC: Proceedings of the International Conference for High Performance Computing, Networking, Storage and Analysis, pp. 1–12 (2008). https://dl.acm.org/doi/abs/10.5555/1413370.1413408
10. Hristova, D., Williams, M.J., Musolesi, M., Panzarasa, P., Mascolo, C.: Measuring urban social diversity using interconnected geo-social networks. In: Proceedings of the 25th International Conference on World Wide Web - WWW '16, pp. 21–30 (2016)

11. CIESIN and CIAT. Gridded population of the world version 3 (GPWv3): Population density grids. Socioeconomic data and applications center (SEDAC). Internet (2019). Accessed 24 Aug 2019
12. Kravari, K., Bassiliades, N.: A survey of agent platforms. J. Artif. Soc. Soc. Simul. **18**(1), 11 (2015)
13. Abar, S., Theodoropoulos, G.K., Lemarinier, P., O'Hare, G.M.P.: Agent based modelling and simulation tools: a review of the state-of-art software. Comput. Sci. Rev. **24**, 13–33 (2017)
14. Rousset, A., Herrmann, B., Lang, C., Philippe, L.: A survey on parallel and distributed multi-agent systems for high performance computing simulations. Comput. Sci. Rev. **22**, 27–46 (2016)
15. Kivela, M., Arenas, A., Barthelemy, M., Gleeson, J.P., Moreno, Y., Porter, M.A.: Multilayer networks. J. Complex Netw. **2**(3), 203–271 (2014)
16. Macal, C., North, M., Pieper, G., Drugan, C.: Modeling: agent-based modeling and simulation for EXASCALE computing. SciDAC Rev. **8**, 34–41 (2008)
17. Luke, S., Cioffi-Revilla, C., Panait, L., Sullivan, K., Balan, G.: Mason: a multiagent simulation environment. Simulation **81**(7), 517–527 (2005)
18. Richmond, P., Walker, D., Coakley, S., Romano, D.: High performance cellular level agent-based simulation with flame for the GPU. Brief. Bioinform. **11**(3), 334–347 (2010)
19. Rubio-Campillo, X.: Pandora: a versatile agent-based modelling platform for social simulation. In: The Sixth International Conference on Advances in System Simulation, SIMUL 2014, pp. 29–34 (2014)
20. Cordasco, G., Spagnuolo, C., Scarano, V.: Work partitioning on parallel and distributed agent-based simulation. In: 2017 IEEE International Parallel and Distributed Processing Symposium Workshops (IPDPSW), pp. 1472–1481 (2017)
21. Barnes, J., Hut, P.: A hierarchical $O(N \log N)$ force-calculation algorithm. Nature **324**(6096), 446–449 (1986)
22. Leskovec, J., Sosič, R.: Snap: a general-purpose network analysis and graph-mining library. ACM Trans. Intell. Syst. Technol. **8**(1), 1–20 (2016)
23. Gonzalez, J.E., Low, Y., Gu, H., Bickson, D., Guestrin, C.: Powergraph: distributed graph-parallel computation on natural graphs. In: Presented as part of the 10th USENIX Symposium on Operating Systems Design and Implementation (OSDI 12), pp. 17–30. USENIX, Hollywood (2012)
24. Axelrod, R.: The dissemination of culture: a model with local convergence and global polarization. J. Conf. Resolut. **41**(2), 203–226 (1997)
25. Castellano, C., Fortunato, S., Loreto, V.: Variants of the Axelrod mode. in statistical physics of social dynamics. Rev. Mod. Phys. **81**(2), 591–646 (2009)
26. Lanchier, N.: The Axelrod model for the dissemination of culture revisited. Ann. Appl. Probab. **22**(2), 860–880 (2012)

Affecting the Relaxation Parameter in the Multifrontal Method

Tomoki Nakano, Mitsuo Yokokawa, Yusaku Yamamoto, and Takeshi Fukaya

Abstract Numerical solutions of a large sparse linear system of equations are often appeared in numerical simulations. In many cases, the computational time to solve it accounts for a large portion of the total simulation time. Thus, reducing its time is very important. We studied a relaxed supernodal multifrontal method for the direct method of symmetric positive definite linear systems, and numerical experiments on test matrices from the University of Florida Sparse Matrix Collection are presented. We implemented two codes. One is naive implementation which is called basic code. The other is enhanced code which is modified from the basic code in terms of reducing the number of data movement of frontal and update matrices. We found two facts. Firstly, enhanced code is better than basic code in terms of memory storage. Secondly, the performance of this method depends on a relaxation parameter which coalesces the supernodes. Furthermore, this optimal parameter depends on matrices, detailed implementation, and machine architecture.

1 Introduction

Numerical solutions of a large sparse linear system of equations are often appeared in numerical simulations. Such systems arise mainly from discrete approximations of partial differential equations. In many cases, the time to solve it accounts for a large portion of the total simulation time. Thus, reducing its time is very important. There are two kinds of methods, direct methods and iterative methods. There is a vast variety in both methods. Whether each method is good or bad depends on size

T. Nakano (✉) · M. Yokokawa
Kobe University, Nada-ku, Kobe, Japan
e-mail: tomoki_nakano@stu.kobe-u.ac.jp

Y. Yamamoto
The University of Electro-Communications, Chofu, Tokyo, Japan

T. Fukaya
Hokkaido University, Sapporo, Hokkaido, Japan

© Springer Nature Switzerland AG 2020
M. M. Resch et al. (eds.), *Sustained Simulation Performance 2018 and 2019*,
https://doi.org/10.1007/978-3-030-39181-2_17

and property of matrices, and problem setting. Considering them, we must select the optimal solution for the simulation. We studied direct methods for SPD (symmetric positive definite) sparse linear systems [1–3].

The direct methods are suited for problems in which the coefficient matrix doesn't change during simulations, but right-hand side of a given system changes. A procedure of solving $A\mathbf{x} = \mathbf{b}$ is decomposed into three steps:

1. a coefficient matrix A is factorized into the two factors LU, where L is a lower triangular matrix and U is a upper triangular matrix,
2. a triangular system $L\mathbf{y} = \mathbf{b}$ is solved, and,
3. a triangular system $U\mathbf{x} = \mathbf{y}$, is solved.

Step 1 is usually the most time-consuming part, whereas step 2 and step 3 are about an order of magnitude faster. Moreover, step 1 doesn't need to be applied more than twice, where the coefficient matrix doesn't change through the simulation. Therefore, this problem is solved efficiently.

2 Numerical Method

2.1 Sparse Solver

In this section, we describe an overview of the sparse solver for SPD linear systems. When the matrix is SPD, it can be factorized into $A = LL^T$ with no pivoting by the *Cholesky factorization*. The solver is decomposed three phases: ANALYZE, FACTORIZE and SOLVE. These phases are summarized in Fig. 1. ANALYZE phase works on the sparsity pattern alone and involves no actual computation on real numbers. FACTORIZE phase works on the coefficient matrix alone and is independent of the right-hand side.

Fig. 1 Overview of the direct sparse solver. Where the nonzero pattern of the matrix doesn't change, ANALYZE phase needs not to be performed again (A). When the matrix doesn't change, ANALYZE and FACTORIZE phases don't need to be performed again (B)

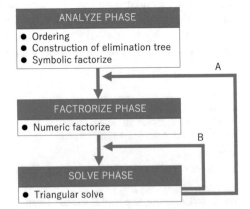

The direct methods for a linear system with a sparse matrix consider the *fill-in* which is the introduction of new nonzeros in the factors that do not appear in the corresponding positions in the matrix being factorized. The nonzero pattern of the Cholesky factor L is determined by *symbolic factorization* which is carried out before the actual numerical factorization. The various phases use an *elimination tree* which provides information relevant to the sparse matrix factorization. The elimination tree associated with the Cholesky factor $L = (l_{ij})$ is the tree such that a parent node of each node j is $parent(j)$ where

$$parent(j) = \min\{i : i > j \wedge l_{ij} \neq 0\}. \tag{1}$$

It is possible to find a permutation matrix P before factorizing, and solve a reordered system

$$(PAP^{\mathrm{T}})(Px) = Pb. \tag{2}$$

Good permutation reduces the number of fill-in elements, computational complexity, or memory storage.

2.2   Multifrontal Method

The *multifrontal method*[4, 5] recognizes the overall factorization of a sparse matrix into a sequence of partial factorizations of dense smaller matrices. The process to eliminate column j is as follows:

1. assemble the frontal matrix F_j from a_{*j} and the update matrices U_i for $i \in C_j$ where C_j is a set of j's child nodes in the elimination tree,
2. perform one step elimination for F_j, and
3. strip off the first column of F_j and store l_{*j}, leaving the update matrix U_j.

The assembly process can be done with indexed vector operations using local indices[6], basically an _AXPYI operation which adds a scalar multiple of compressed sparse vector to a full-storage vector.

This method can use Level 2 BLAS because the elimination is performed on a dense matrix. However, it suffers from a disadvantage of allocating a working storage for frontal and update matrices, and copying the update and frontal matrices to and from the working storage.

Sequential multifrontal method is performed in postorder on the tree. In the context of this method, it has the following additional desirable property: the update matrices can be managed in a last-in-first-out basis. Therefore, we can manipulate the sequence of update matrices using a stack. Liu[7] estimates the size of working storage. In addition, he minimizes this size by reordering, but we don't apply this technique in this paper. Algorithm 1 captures the essence of the sequential

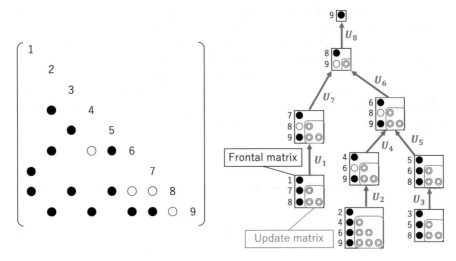

Fig. 2 An example of the frontal and update matrices when performing the multifrontal method. Black rectangles are denoted frontal matrices, and blue rounded rectangles are denoted update matrices. Filled circle are nonzeros in A and open circle are fill-ins

multifrontal method. $\oplus$ is to assemble the frontal matrix and the update matrix. Figure 2 shows an example of the multifrontal method.

Algorithm 1: Sequential multifrontal method

Input: The SPD matrix $A = (a_{ij})$ which is postordered.
$\quad\quad$ C_j is the set of j's children on elimination tree.
Output: The Cholesky factor $L = (l_{ij})$
for $j := 1$ **to** n **do**
$\quad$ Set a_{*j} to F_j
$\quad$ **foreach** $i \in C_j$ **do**
$\quad\quad$ Pop an update matrix U_i from the stack
$\quad\quad$ $F_j := F_j \oplus U_i$
$\quad$ **end foreach**
$\quad$ Perform 1 step of elimination on F_j to give the factor columns l_{*j} and U_j
$\quad$ Push the update matrix U_j into the stack
end for

2.3 Supernode

A practical improvement to the multifrontal method is the use of *supernodes*. Generally speaking, columns are grouped together to form a supernode if they can be treated as one computational unit in the course of sparse Cholesky factorization. A supernode is a maximal block of contiguous columns where the diagonal block is full triangular matrix, and these columns all have identical off-block-diagonal column structures. The *supernodal elimination tree* is defined to be the tree consisting of the supernodes. A *fundamental supernode* is a supernode such that each column has only child in the elimination tree. Liu et al. [8] obtain the fundamental supernode by $\mathcal{O}(\eta(A))$ where $\eta(A)$ is the number of nonzeros in A.

2.4 Supernodal Multifrontal Method

All columns in a supernode share the same frontal matrix and are eliminated together. The total number of matrix assemblies during the factorization is reduced from the number of columns to the number of supernodes. The process for eliminating the p columns of a supernode $J = (j_1, \ldots, j_p)$ is as follows:

1. to assemble the frontal matrix F_J from a_{*j} for $j \in J$ and the update matrices U_K for supernodes $K \in \mathcal{C}_J$ where $\mathcal{C}_J$ is a set of J's child supernodes in the supernodal elimination tree,
2. to perform p steps elimination for F_J, and
3. to strip off the first p columns of F_j and store l_{*j} for $j \in J$, leaving the update matrix U_J.

The supernodal multifrontal method has two advantages compared with the (nodal) multifrontal method. One advantage is that it can use Level 3 BLAS because of performing multi-step elimination. The other one is that the number of copying update matrices from and to stack and the number of _AXPYI operations decrease.

2.5 Relaxed Supernode

There are computational advantages in having large supernodes. Ashcraft and Grimes [9] suggest the amalgamation of nodes/supernodes to allow for larger supernodes.

The decision on whether two supernodes are merged or not depends on the number of zero entries introduced into the factorization for the merged supernode, including any already existing zeros introduced in the two supernodes from previous merges. We introduced the parameter, *max_zero*, to represent the number of allowable zero entries. The algorithm is as follows. When each supernode encountered in a postorder traversal of the fundamental supernodal tree, the algorithm first determines the subset of sons with each of which it can be merged without generating a supernode with more than *max_zero* zeros in its data structure. It then merges the son that adds the fewest number of zeros with the current supernode. This process ensures that once a son is merged into a supernode, its sons (grandsons of the supernode) do not need to be tested for a possible merge. If a supernode cannot merge with its parent, it cannot merge with its grandparent when the grandparent has absorbed the parent.

2.6 Enhancements to the Supernodal Multifrontal Method

We use the following two techniques in [10] which are some enhancements to the basic multifrontal method to implement our code. Firstly, their method compute the new factor columns $L_{*,J}$ in factor storage rather than in F_J and the compute only the update matrix U_J within the frontal matrix F_J. This reduces the size of the working storage and eliminates the need to move eliminated columns from the frontal matrix to factor storage.

Secondly, the columns of U_J destined for $L_{*,P}$ are assembled directly into factor storage by using local indices and remainder of the update matrix only copy to the stack. This has the same advantages as the first technique. In [10], this technique doesn't use only parent node, but also ancestor nodes. However, our code uses only the parent node.

3 Performance Evaluation

In this section we compare the performance for using the different max_zero parameters. In addition, we compare the working storage and time of the following two multifrontal solvers:

1. `basic-mf`: basic multifrontal method, and
2. `enhanced-mf`: enhanced multifrontal method described in Sect. 2.6.

We conducted experiments on two systems, Intel Xeon Phi 7250 (Knights Landing) and Intel Xeon E7-8857 (Ivy-Bridge). We use the following two libraries:

1. METIS: nested dissection ordering for fill-in reduction[11], and
2. MKL BLAS: high performance matrix operation.

Our solvers were coded in Fortran2003, and all floating-point operations were performed in double precision. These were compiled by *Intel Fortran Compiler* with "*-fast*" option for optimization on Xeon and Xeon Phi. "*-fast*" option is equivalent to "*-xHOST -O3 -ipo -no-prec-div -static -fp-model fast=2*".

3.1 Comparison of Required Working Storage

Several matrices from the University of Florida Sparse Matrix Collection [12] were selected for the experiments and some of their characteristics are provided in Table 1. Op. count in the fifth column in the table was obtained from the column counts[1]:

$$\sum_{j=1}^{n} |\mathcal{L}_j|^2 \tag{3}$$

Table 1 Characteristics of test matrices and the size of workstorages

Matrix	n	$\|A\|$	$\|L\|$	Op. count	Wk-basic	Wk-enhanced
apache1	80,800	542,184	10,245,999	5,499,074,179	2,734,497	1,937,429
bcsstk36	23,052	1,143,140	2,718,323	531,411,213	511,252	237,310
cfd1	70,656	1,825,580	19,236,449	10,835,799,583	3,561,677	1,882,160
gridgena	48,962	512,084	2,618,163	341,552,705	346,316	182,900
gyro	17,361	1,021,159	1,520,005	191,016,959	120,597	93,330
Pres_poisson	14,822	715,804	2,353,485	521,457,075	466,396	211,636
qa8fm	66,127	1,660,579	21,374,928	17,737,522,598	5,917,029	2,958,525
raefsky4	19,779	1,229,776	5,405,333	2,293,934,565	1,480,754	859,217
shallow_water1	81,920	327,680	1,961,034	306,449,950	529,600	405,007
vanbody	47,072	2,329,056	5,687,111	1,215,984,803	693,734	479,668

n: number of equations; $\|A\|$: number of nonzeros in A; $\|L\|$: number of nonzeros in L; Op. count: number of floating-point operations required to compute L; Wk-basic: array size of required working storage for `basic-mf`; Wk-enhanced: array size of required working storage for `enhanced-mf`

where $\mathcal{L}_j$ is a set of the row indices of nonzero element in column j. Table 1 also shows the size of required working storages for `basic-mf` and `enhanced-mf`. For all matrices, the required size of working storage in `enhanced-mf` is smaller than in `basic-mf`. The most reduction in the required size of working storage of `enhanced-mf` over `basic-mf` is 53% for the matrix bcsstk36.

3.2 Performance on Xeon Phi

The specification of Xeon Phi shows in Table 2[13]. This machine has DDR4 memories and MCDRAM memories. User selects the following three memory modes:

1. cache mode: MCDRAM is a L3 cache for DDR,
2. flat mode: MCDRAM is treated like standard memory, and
3. hybrid mode: a portion of MCDRAM is L3 cache and the remainder is flat.

We select the flat mode because required the size of our code for any of matrices is smaller than the sum of MCDRAM memories. User can also select the three cluster modes: all-to-all, quadrant and SNC mode. We select all-to-all mode.

Table 3 shows the factorization times on Xeon Phi with `basic-mf`. The optimal *max_zero* with respect to the factorization time range from 2^{10} to 2^{14}. The speedup over the *max_zero* $= 0$ case ranges from 1.28 to 3.29 times for all matrices. Table 4 shows factorization times on Xeon Phi with `enhanced-mf`. The optimal *max_zero* with respect to the factorization time range from 2^{11} to

Table 2 The specification of Intel Xeon Phi 7250

Code name	Knights landing
Number of clocks	1.40 GHz
Number of cores	68
L1 instruction cache	32 KB
L1 data cache	32 KB
Distributed L2 cache	1 MB × 34
FLOPS/Clock	32
Theoretical peak operation performance	3.05 TFlops
DDR4 memory	64 GB × 6 (90 GB/s)
MCDRAM	2 GB × 8 (400 GB/s)

Table 3 Influence of *max_zero*: factorization times in seconds on Xeon Phi (`basic-mf`)

max_zero	0	2^7	2^8	2^9	2^{10}	2^{11}	2^{12}	2^{13}	2^{14}	2^{15}
apache1	1.633	1.000	0.936	0.873	0.845	0.831	**0.826**	0.828	0.872	0.971
bcsstk36	0.237	0.209	0.201	0.189	0.189	0.186	**0.185**	0.186	0.203	0.228
cfd1	1.795	1.407	1.366	1.333	1.300	1.269	1.248	**1.214**	1.223	1.285
gridgena	0.527	0.256	0.240	0.221	**0.216**	0.218	0.225	0.255	0.291	0.339
gyro	0.177	0.129	0.123	0.111	0.113	**0.106**	0.110	0.113	0.129	0.144
Pres_poisson	0.233	0.161	0.157	0.154	0.147	0.143	**0.140**	0.140	0.149	0.158
qa8fm	2.031	1.654	1.629	1.591	1.578	1.543	1.550	**1.499**	1.546	1.565
raefsky4	0.492	0.416	0.396	0.380	0.367	0.362	0.345	0.338	**0.329**	0.345
shallow_water1	0.848	0.316	0.288	0.264	**0.258**	0.267	0.296	0.335	0.428	0.502
vanbody	0.540	0.439	0.409	0.395	0.381	**0.379**	0.380	0.389	0.417	0.484

The bold number is the fastest in each matrix

Table 4 Influence of *max_zero*: factorization times in seconds on Xeon Phi (`enhanced-mf`)

max_zero	0	2^7	2^8	2^9	2^{10}	2^{11}	2^{12}	2^{13}	2^{14}	2^{15}
apache1	1.588	0.947	0.877	0.796	0.765	0.743	0.728	**0.725**	0.767	0.852
bcsstk36	0.220	0.188	0.182	0.169	0.165	0.160	**0.154**	0.159	0.173	0.193
cfd1	1.702	1.299	1.251	1.218	1.187	1.155	1.122	**1.097**	1.106	1.181
gridgena	0.512	0.243	0.218	0.191	0.184	**0.181**	0.182	0.202	0.233	0.277
gyro	0.165	0.118	0.110	0.100	0.097	0.092	**0.091**	0.097	0.109	0.122
Pres_poisson	0.205	0.143	0.137	0.137	0.130	0.121	**0.120**	0.121	0.123	0.136
qa8fm	1.922	1.563	1.525	1.491	1.468	1.444	1.412	1.392	**1.392**	1.432
raefsky4	0.456	0.379	0.361	0.348	0.330	0.334	0.317	0.305	**0.295**	0.311
shallow_water1	0.850	0.296	0.266	0.245	0.237	**0.234**	0.247	0.270	0.341	0.398
vanbody	0.506	0.402	0.373	0.351	0.331	0.322	**0.321**	0.331	0.358	0.403

The bold number is the fastest in each matrix

2^{14}. The speedup over the *max_zero* $= 0$ case ranges from 1.38 to 3.63 times for all matrices. Except for the matrix shallow_water1 where *max_zero* $= 0$, `enhanced-mf` is faster than `basic-mf`. However, the optimal *max_zero* is different in `basic-mf` and `enhanced-mf` for some matrices. This turns out that the optimal max_zero doesn't only depend on a matrix, but also detailed implementation.

3.3 Performance on Xeon

The specification of Xeon shows in Table 5. Table 6 shows factorization times on Xeon with `enhanced-mf`. The optimal max_zero with respect to the factorization time range from 2^8 to 2^{10}. The speedup over the $max\_zero = 0$ case ranges from 1.06 to 1.72 times for all matrices. The optimal max_zero on Xeon is smaller than that on KNL for all matrices. It looks that the difference of the ratio of the speed of dense vector operations to sparse vector operations between two processors. The number of floating point operations per clock is 8 i.e. 4-wide AVX addition + 4-wide AVX multiplication for Xeon. On the other hand, it is 32 i.e. two 8-wide FMA instructions for Xeon Phi. We found out that the optimal max_zero depends on machine architecture.

Table 5 The specification of Intel Xeon E7-8857

Code name	Ivy-Bridge
Number of clocks	3.0 GHz
Number of cores	12
L1 instruction cache	32 KB
L1 data cache	32 KB
L2 cache	256 KB
L3 cache	30 MB
FLOPS/Clock	8
Theoretical peak operation performance	288 GFlops
DDR3 memory	32GiB * 512

Table 6 Influence of max_zero: factorization times in seconds on Xeon (`enhanced-mf`)

max_zero	0	2^7	2^8	2^9	2^{10}	2^{11}	2^{12}	2^{13}	2^{14}	2^{15}
apache1	0.598	0.489	0.484	**0.476**	0.478	0.483	0.500	0.534	0.599	0.711
bcsstk36	0.084	0.078	0.077	**0.075**	0.076	0.079	0.086	0.098	0.115	0.143
cfd1	0.932	0.860	0.851	0.847	**0.845**	0.847	0.861	0.891	0.957	1.063
gridgena	0.104	0.073	**0.071**	0.071	0.074	0.082	0.094	0.115	0.146	0.187
gyro	0.050	0.040	0.040	0.039	**0.039**	0.041	0.046	0.053	0.067	0.083
Pres_poisson	0.080	0.065	0.064	**0.063**	**0.063**	0.065	0.068	0.075	0.083	0.102
qa8fm	1.289	1.222	1.217	**1.213**	1.214	1.216	1.228	1.254	1.310	1.417
raefsky4	0.231	0.212	0.209	0.206	**0.204**	0.206	0.208	0.219	0.233	0.265
shallow_water1	0.117	0.068	**0.069**	0.072	0.077	0.090	0.115	0.140	0.197	0.243
vanbody	0.192	0.169	0.165	**0.162**	0.163	0.171	0.182	0.206	0.243	0.304

The bold number is the fastest in each matrix

4 Conclusion

We introduced the supernodal multifrontal method with relaxing supernodes, and some enhanced techniques. On Xeon Phi and Xeon, we conduct performance evaluations with the parameter, max_zero and compare the performance between basic-mf and enhanced-mf. We have found the two facts. Firstly, enhanced-mf can save the working storage compared to basic-mf. Secondly, the optimal max_zero depends on solving matrix, a implementation, and a used machine. From this fact, predicting the optimal max_zero is very difficult in advance of calculation. One possible solution for this problem is automatic tuning. This is the one of the future works. Another future work is parallelization of our code for the multicore architecture.

Acknowledgement This work was supported in part by JSPS KAKENNHI Grant Number (B)17H02828 and (C)18K11325.

References

1. Davis, T.A., Rajamanickam, S., Sid-Lakhdar, W.M.: A survey of direct methods for sparse linear systems. Acta Numer. **25**, 383–566 (2016). https://doi.org/10.1017/S0962492916000076
2. Davis, T.A.: Direct Methods for Sparse Linear Systems. SIAM, Philadelphia (2006)
3. Duff, I.S., Erisman, A.M., Reid, J.K.: Direct Methods for Sparse Matrices. Oxford University Press, Oxford (2017)
4. Duff, I.S., Reid, J.K.: The multifrontal solution of indefinite sparse symmetric linear. ACM Trans. Mathe. Softw. **9**(3), 302–325 (1983)
5. Liu, J.W.: The multifrontal method for sparse matrix solution: theory and practice. SIAM Rev. **34**(1), 82–109 (1992)
6. Schreiber, R.: A new implementation of sparse Gaussian elimination. ACM Trans. Math. Softw. **8**(3), 256–276 (1982)
7. Liu, J.W.H.: On the storage requirement in the out-of-core multifrontal method for sparse factorization. ACM Trans. Math. Softw. **12**(3), 249–264 (1986). https://doi.org/10.1145/7921.11325
8. Liu, J.W., Ng, E.G., Peyton, B.W.: On finding supernodes for sparse matrix computations. SIAM J. Matrix Anal. Appl. **14**(1), 242–252 (1993)
9. Ashcraft, C., Grimes, R.: The influence of relaxed supernode partitions on the multifrontal method. ACM Trans. Math. Softw. **15**(4), 291–309 (1989). https://doi.org/10.1145/76909.76910
10. Ng, E.G., Peyton, B.W.: Block sparse Cholesky algorithms on advanced uniprocessor computers. SIAM J. Sci. Comput. **14**(5), 1034–1056 (1993)
11. Karypis, G., Kumar, V.: A fast and high quality multilevel scheme for partitioning irregular graphs. SIAM J. Sci. Comput. **20**(1), 359–392 (1998)
12. Davis, T.A., Hu, Y.: The University of Florida sparse matrix collection. ACM Trans. Math. Softw. **38**(1), 1–25 (2011)
13. Sodani, A., Gramunt, R., Corbal, J., Kim, H.S., Vinod, K., Chinthamani, S., Hutsell, S., Agarwal, R., Liu, Y.C.: Knights landing: second-generation Intel Xeon Phi product. IEEE Micro **36**(2), 34–46 (2016)

Enhancement of the *GW* Space-Time Program Code for Accurate Prediction of the Electronic Properties of Organic Electronics Materials

Susumu Yanagisawa, Takeshi Yamashita, and Ryusuke Egawa

Abstract For large-scale efficiently parallelized electronic structure calculation within the *GW* approximation, we modified the MPI-parallelized version of the *GW* space-time program. To reduce the communication time required for computation of the inverse of the complex dielectric matrix, which is one of the bottlenecks of the program, the ScaLapack library codes employed for the LU-decomposition matrix inversion was replaced with the Lapack counterpart implemented with the intranode task parallelization. As a result, the elapsed time for matrix inversion significantly reduced, along with improvement on the parallelization efficiency for the number of nodes or cores. In addition, the intranode task parallelization for inversion with OpenMP was found to show reasonable parallelization efficiency with respect to the number of threads inside a node. Overall, the improvement in computation time will allow us to investigate not only the electronic structure of bulk phases, but also those of surfaces and interfaces of organic molecular crystals.

1 Introduction

Electronics device materials comprised of organic molecular solids have attracted considerable attention as the next-generation flexible electronics because of the advances such as their low-cost printing-like fabrication process and low power

S. Yanagisawa (✉)
Department of Physics and Earth Sciences, Faculty of Science, University of the Ryukyus,
Nishihara, Okinawa, Japan
e-mail: shou@sci.u-ryukyu.ac.jp

T. Yamashita
Information Infrastructure Division, Information Department, Tohoku University, Aoba-ku,
Sendai, Japan
e-mail: yamacta@tohoku.ac.jp

R. Egawa
Cyberscience Center, Tohoku University, Aoba-ku, Sendai, Japan
e-mail: egawa@cc.tohoku.ac.jp

© Springer Nature Switzerland AG 2020

M. M. Resch et al. (eds.), *Sustained Simulation Performance 2018 and 2019*,
https://doi.org/10.1007/978-3-030-39181-2_18

consumption. While the organic light-emitting diodes (OLEDs) are already on the market, other devices such as the organic field-effect transistors (OFETs) and the organic photovoltaics (OPVs) still remain to be commercialized. In terms of the basic electronic properties, there is no consensus on the mechanism of the essential electronic phenomena such as the carrier transport. For insights into the electronic origins of the phenomena, roles of theoretical electronic structure simulations, specifically the first-principles electronic structure methods, which do not depend on specific experimental values or on empirical models, are becoming important.

To play a role in prediction or elucidation of the electronic properties such as the barrier for carrier injection and the mobility of a carrier, quantitative accuracy beyond the prevailing approximations to the density functional theory (DFTA) is being required. These years, many-body perturbation theory within the GW approximation[1, 2] has been a method of choice in prediction of the (quasiparticle) energy levels of electrons and holes that are injected into organic molecular solids. Its predictive power has been demonstrated for the fundamental band gap, band structure, or barrier for charge injection of organic semiconductor crystals[3–9], which has led us to insights into the induced polarization or the screening effect upon the injected charge and the role of the electrostatic effects depending on the molecular orientation at the surface.

The energy level alignment at organic-metal interface, one of the problems of crucial importance dominating the barrier for charge injection from the metal electrode to the organic layer, can be described well by the GW approximation because of the inclusion of the screening effect induced by the image charge at the conducting metal substrate[3, 4]. Using the same method, the long-ranged image potential states at the metal surface can be correctly reproduced as well. However, there have been a few publications reporting the result on the application of the GW to describing electronic structures of organic semiconductor surfaces or organic-metal interfaces[3, 4, 10–13]. One of the main reasons of the limited applications seems to be the large computational resources required, whose computational demand increases on the other of N^4, where N is the number of atoms in the unit cell, or the number of basis set[14].

In this study, to extend the applicability of the GW approximation, we promoted computational efficiency of the GW space-time (GWST) program[15–17]. The program code was applied to theoretical determination of the fundamental gap or the ionization energy/electron affinity of typical organic semiconductor crystals[6, 9, 18–21]. Here, we significantly decreased the elapsed time by replacing the ScaLapack library codes with the Lapack library ones in an inversion of the complex dielectric matrix with the LU-decomposition. In addition, we employed the intranode task parallelization of the matrix inversion with OpenMP. We found that the parallelization efficiency increased reasonably.

2 Computational Method

2.1 GWST Method

The detailed basic formalism and the physical background of the GWST method is described elsewhere[15–17]. Here, we make a short summary of the methodology. In the GWST method, one starts from the non-interacting one-body Green's function $G(\mathbf{r}, \mathbf{r}'; i\tau)$ in real space $(\mathbf{r}, \mathbf{r}')$ and in imaginary time $(i\tau)$ for propagation of the hole $(\tau > 0)$ and the electron $(\tau < 0)$, respectively,

$$
G(\mathbf{r}, \mathbf{r}'; i\tau) = \begin{cases} i \sum_{n\mathbf{k}}^{\text{occ}} \phi_{n\mathbf{k}}(\mathbf{r}) \phi_{n\mathbf{k}}^*(\mathbf{r}') \exp(\epsilon_{n\mathbf{k}}\tau) & (\tau > 0), \\ -i \sum_{n\mathbf{k}}^{\text{unocc}} \phi_{n\mathbf{k}}(\mathbf{r}) \phi_{n\mathbf{k}}^*(\mathbf{r}') \exp(\epsilon_{n\mathbf{k}}\tau) & (\tau < 0). \end{cases} \tag{1}
$$

The Green's function is constructed from eigenvalues $\epsilon_{n\mathbf{k}}$ and wave functions $\phi_{n\mathbf{k}}(\mathbf{r})$ obtained with a mean-field approximation such as the density functional theory within the local density or a generalized gradient approximation (DFT-LDA or DFT-GGA), and $\mathbf{k}$ vectors denote those in the first Brillouin zone. Within the random-phase approximation (RPA), the irreducible polarization in real space and imaginary time is obtained as

$$
P(\mathbf{r}, \mathbf{r}'; i\tau) = -2i G(\mathbf{r}, \mathbf{r}'; i\tau) G(\mathbf{r}', \mathbf{r}; -i\tau), \tag{2}
$$

which is Fourier transformed to $P_{\mathbf{GG}'}(\mathbf{k}; i\omega)$ in reciprocal space and imaginary frequency. The symmetrized dielectric matrix in reciprocal space $\tilde{\epsilon}_{\mathbf{GG}'}(\mathbf{k}; i\omega)$ is described as

$$
\tilde{\epsilon}_{\mathbf{GG}'}(\mathbf{k}; i\omega) = \delta_{\mathbf{GG}'} - \frac{4\pi}{|\mathbf{k}+\mathbf{G}||\mathbf{k}+\mathbf{G}'|} P_{\mathbf{GG}'}(\mathbf{k}; i\omega). \tag{3}
$$

Then, one obtains the screened Coulomb potential $W(\mathbf{r}, \mathbf{r}'; i\tau)$ by Fourier transforming to real space and imaginary time the screened potential in reciprocal space and imaginary frequency,

$$
W_{\mathbf{GG}'}(\mathbf{k}; i\omega) = \frac{4\pi}{|\mathbf{k}+\mathbf{G}||\mathbf{k}+\mathbf{G}'|} \tilde{\epsilon}_{\mathbf{GG}'}^{-1}(\mathbf{k}; i\omega). \tag{4}
$$

Finally, the self-energy operator is calculated by the product of the screened potential and the Green's function,

$$
\Sigma(\mathbf{r}, \mathbf{r}'; i\tau) = i G(\mathbf{r}, \mathbf{r}'; i\tau) W(\mathbf{r}, \mathbf{r}'; i\tau). \tag{5}
$$

Notice that the quantities in Eqs. 2–5 are obtained by simple multiplications, without convolution in the frequency domain as required in the conventional methodology of the *GW* approximation[15]. To retain simple multiplication without convolution throughout the codes, the arguments of the calculated quantities in Eqs. 1–5, i.e., real space and reciprocal space vectors, and imaginary time and imaginary frequencies, are switched by the Fast Fourier Transformations (FFTs)[15].

The scaling in the conventional GW calculation involving convolution is the fourth power of the number of the reciprocal space grid (N_G) and quadratic with that of the frequency grid. The number of multiplications required in Eqs. 2–5 of GWST in general scales quadratically in the number of the real space grid (N_r) or N_G, and linearly with respect to the number of the imaginary time grid (N_τ) or the imaginary frequency grid (N_ω), thus implying an advantage in computational cost over the conventional method. The sum over states involving the multiplication as shown in Eq. 1 can be efficiently manipulated as a matrix product by using source codes in Level 3 BLAS library[17].

2.2 Modification on the Program Code: Parallelization in Inversion of the Dielectric Matrix

As described in Eq. 4, the inverse of the complex dielectric matrix is required. For parallel algorithms, the matrices are distributed over the processing elements (PEs). Parallelized inversion of the complex dielectric matrix, which is Hermitian, and is in general not sparse, involved ScaLapack library codes for matrix inversion with the LU decomposition[14]. There is a concern, however, on a possible decrease in parallel efficiency during the computation, i.e. the time required for communication between the PEs. We performed test calculations with or without ScaLapack library codes, so as to check the scaling of the matrix inversion and the overall performance of the program.

To facilitate comparison between the ScaLapack library-based inversion and the non-ScaLapack inversion in terms of the communication time, we implemented a code in which the Lapack library codes for the LU decomposition and inversion was employed locally on each of the PEs. Given Eq. 4, the inversion can be performed independently on each of the PEs by distributing the dielectric matrices over the PEs with respect to **k**–points in the Brillouin zone and imaginary frequencies $i\omega$ (for schematics, see Fig. 1). In the test calculation (naphthalene crystal; see Fig. 2), the 14 **k**–points and 16 imaginary frequency grids were used, for which sufficient convergence of the calculated quasiparticle energies was confirmed in the previous study[21]. Overall, we employed 224 PEs for MPI parallelization, and one core per node was used on NEC SX-ACE. The number of the plane wave basis set or **G**-vectors, which is equal to the rank of the dielectric matrix $\epsilon_{\mathbf{GG'}}$, was set to 8279, 14,965, and 20,121, that corresponded to cutoff for the kinetic energy of plane waves of 36 Ry, 54 Ry, and 65 Ry, respectively.

In addition, we checked the performance of the intranode task parallelization for matrix inversion with OpenMP. To facilitate the OpenMP task parallelization, we employed the Lapack library code as implemented in the MATHKEISAN numerical

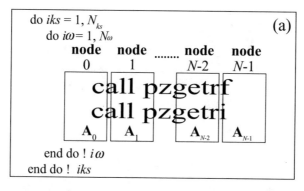

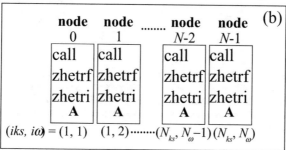

Fig. 1 Schematics of the parallelized matrix inversion of the complex dielectric matrix used in this work. (**a**) Use of the ScaLapack library codes. Notice that the submatrices distributed over all the nodes or the processing elements (PEs) are denoted by A_i. (**b**) The modified code without the ScaLapack library codes. The dielectric matrix to be inverted of each of the (**k**-point, $i\omega$) sets is assigned to each of the node or the PEs. On each of the nodes or PEs, the inversion is locally done with the Lapack library codes. With the codes allowing OpenMP parallelization, the intranode parallelization is also available (see text)

library, so that up to four threads were used for the task parallelization in a node of SX-ACE.

3   Results and Discussion

3.1   Performance of GWST Program Employing Complex Matrix Inversion with the ScaLapack Library Code

Table 1 displays the overall performance of GWST using the ScaLapack library code for the LU-composition and the following inversion of the dielectric matrices, showing that the overall elapsed time does not scale linearly even when the number of cores exceeds 50. Increase in proportion of the elapsed time for inversion relative to the overall elapsed time in accordance with increase in the number of PEs

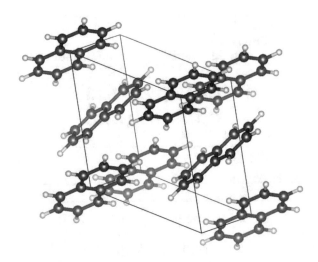

Fig. 2 Overview of a naphthalene single crystal. The crystal unit cell consists of two molecules, and solid lines display its edges. The smaller and larger spheres denote H and C atoms, respectively. Some periodic replicas of the molecules are also displayed. The cell lengths and angles of the unit cell and the atomic configurations come from those predicted by a van der Waals-inclusive first-principles calculation [21]

Table 1 The overall elapsed time of the program code for GW calculation of naphthalene crystal with GWST program code

Number of PEs (cores)	Overall elapsed time (s)	Mat. Inv. (s)	Proportion (%)
32	10, 836	1315	12.1
64	6061	995	16.4
128	3716	788	18.8
256	2554	702	27.5
512	1995	657	32.9

The third and the fourth columns indicate the elapsed time spent on inversion and its proportion relative to the overall elapsed time, respectively. Notice that the elapsed time includes time spent for communication between the processing elements. The rank of the complex dielectric matrix is 8279. All the timing data were obtained with the performance analysis tool "FTRACE" as implemented in NEC SX-ACE

(see third and fourth column in Table 1) implies that the low parallel efficiency of the inversion with the ScaLapack library code affects the overall performance of GWST. It is also found that the low parallel efficiency of the inversion comes from its communication time required for the transfer of data between the PEs. The communication time in the computation with 128–512 PEs covers 86–95% and 58–74% of the elapsed time in the LU-decomposition (PZGETRF) and the following inversion (PZGETRI), respectively. In case the rank of the complex dielectric matrix changed into 20,121, the large portion of the communication relative to the overall elapsed time retained. The situation might improve when a rank of the dielectric matrix larger by one order or more was taken into account. However, such a huge sized matrix might not be treated by the program code at present, and, we thus would like to leave it to future work to examine the performance for a dielectric matrix with a higher rank. Nevertheless, notice that moderate energy cutoff or the number of reciprocal space **G** vectors for polarizability or dielectric matrix is required for organic molecular crystals in general[6, 22], and thus the performance test on the matrix inversion with its rank of the order similar to the present work is required. Next, we will describe possible improvement of the program code for the present test set.

3.2 Parallel Efficiency of GWST with a Dielectric Matrix for Each of the k-Point and Imaginary Frequency iω Grids Owned by One PE

Table 2 shows the overall performance comparison of the GWST code with the ScaLapack library code and that without ScaLapack. It was found that the inversion with the non-ScaLapack code resulted in a decrease in overall elapsed time by 8 (15)% for 32 (64) PEs. The effective parallelism and parallel efficiency of the GWST code based on the Amdahl's law without ScaLapack (with ScaLapack) are 99.88% (99.58%) and 92.72% (79.11%), respectively, demonstrating improvement on parallelism. Notice that the elapsed time in the non-ScaLapack inversion of 267 s for 64-core parallelization, which is approximately three quarters of that for 32-core parallelization rather than half of that, is due to 14 **k**-point × 16 imaginary frequency grids, which is exactly divisible by 32, but is not divisible by 64. The present non-ScaLapack inversion code linearly scales, without any communication between PEs (see also Table 3). Performance in larger size test for a dielectric matrix whose rank is 14,965 is displayed in Table 3. Here, the effective parallelism is improved up to 99.93%. The parallel efficiency is 86.77%.

Table 2 Comparison between the overall elapsed time with the GWST program code for naphthalene crystal with ScaLapack or without ScaLapack

Number of PEs (cores)	w/o ScaLapack (s)	With ScaLapack (s)
32	9964 (464)	10,836 (1315)
64	5163 (267)	6061 (995)

The numbers of PEs or cores were 32 and 64, respectively. In parentheses are the elapsed time for inversion of the dielectric matrix with the LU-decomposition. The rank of the complex dielectric matrix is 8279. All the timing data were obtained with the performance analysis tool "FTRACE" as implemented in NEC SX-ACE

Table 3 The overall elapsed time with the GWST program code for naphthalene crystal without ScaLapack, along with the elapsed time for inversion of the dielectric matrix with the LU-decomposition in parentheses

Number of PEs (cores)	w/o ScaLapack (s)
112	8727 (682)
224	4674 (345)

The numbers of PEs or cores were 112 and 224, numbers exactly dividing 14 **k**-point × 16 imaginary frequency points. The rank of the complex dielectric matrix is 14,965. All the timing data were obtained with the performance analysis tool "FTRACE" as implemented in NEC SX-ACE

3.3 Intranode Task Parallelization of Matrix Inversion with OpenMP

Finally, to facilitate the implementation of the MPI/OpenMP hybrid parallelization for higher parallelization efficiency, we investigated the performance of the intranode task parallelization in the inversion of the complex dielectric matrix. Here, the MATHKEISAN Lapack library source codes for double complex matrix inversion (ZGETRF, ZGETRI) as implemented in SX-ACE were task parallelized. Tables 4 and 5 demonstrate the performance of the parallelized matrix inversion for the dielectric matrix with its rank of 14,965 and 20,121, respectively. It is found that task parallelization of a matrix with a larger rank of 20,121 reasonably performs better than that with 14,965. Specifically, in the LU-decomposition with ZHETRF, the two- and four-thread calculations demonstrate acceleration by 1.8 and 3.2 times, respectively, relative to a single thread. On the other hand, however, the inversion with ZHETRI accelerates by only 1.9 times with four threads. More inspection would be required for inversion of a matrix with higher rank.

Table 4 The elapsed time with the GWST program code for naphthalene crystal for the library codes ZHETRF and ZHETRI, corresponding to LU-decomposition and inversion, respectively

	ZHETRF (s)	ZHETRI (s)
One thread	76.1 (1.00)	150.6 (1.00)
Two threads	42.7 (1.78)	134.2 (1.12)
Four threads	25.6 (2.97)	79.8 (1.89)

The numbers of nodes were 224, and up to four threads were employed in each of the nodes. Values in parentheses are the elapsed time with one thread divided by that in question. The rank of the complex dielectric matrix is 14,965. All the timing data were obtained with the performance analysis tool "FTRACE" as implemented in NEC SX-ACE

Table 5 The elapsed time with the GWST program code for naphthalene crystal for the library codes ZHETRF and ZHETRI, corresponding to LU-decomposition and inversion, respectively

	ZHETRF (s)	ZHETRI (s)
One thread	181.3 (1.00)	361.1 (1.00)
Two threads	99.1 (1.83)	234.5 (1.54)
Four threads	57.5 (3.15)	194.9 (1.85)

The numbers of nodes were 224, and up to four threads were employed in each of the nodes. The rank of the complex dielectric matrix is 20,121. The conventions is the same as that in Table 4

4 Conclusions

To enhance the performance of the *GW* space-time code for calculations of larger systems on SX-ACE, we improved the parallelization efficiency of the inversion of the complex dielectric matrix, which is one of the bottlenecks of the program. First, we implemented the inversion code in which each of the matrices is manipulated on each of the processing elements (PEs) without any communication between the PEs during the inversion, as opposed to the ScaLapack library codes. Considerable reduction in communication time resulted in a reduction of the overall elapsed time. For the near-future implementation of the MPI/OpenMP hybrid parallelization, performance of the intranode task parallelization of the matrix inversion was investigated. For a matrix with its rank of 10,000–20,000, the code for the LU-decomposition was reasonably accelerated. However, the task parallelization of the following inversion was less efficient. Overall, the present modification of the codes and implementation of the new features resulted in improvements on the elapsed time, as well as the parallelization efficiency. Nevertheless, further tests of larger scale calculation would be required.

Acknowledgements This work was supported by Grant-in-Aid for Scientific Research (Fund for the Promotion of Joint International Research (Fostering Joint International Research): No. 16KK0115) from the Japan Society for the Promotion of Science, and by "Joint Usage/Research Center for Interdisciplinary Large-scale Information Infrastructures" and "High Performance Computing Infrastructure" in Japan (Project ID: jh180069-NAH). We acknowledge the Cyberscience Center, Tohoku University, for the use of their facilities.

References

1. Hedin, L.: Phys. Rev. **139**, A796 (1965)
2. Hybertsen, M.S., Louie, S.G.: Phys. Rev. B **34**, 5390 (1986)
3. Neaton, J.B., Hybertsen, M.S., Louie, S.G.: Phys. Rev. Lett. **97**, 216405 (2006)
4. Garcia-Lastra, J.M., Rostgaard, C., Rubio, A., Thygesen, K.S.: Phys. Rev. B **80**, 245427 (2009)
5. Refaely-Abramson, S., Sharifzadeh, S., Govind, N., Autschbach, J., Neaton, J.B., Baer, R., Kronik, L.: Phys. Rev. Lett. **109**, 226405 (2012)
6. Yanagisawa, S., Morikawa, Y., Schindlmayr, A.: Phys. Rev. B **88**, 115438 (2013)
7. Li, J., D'Avino, G., Duchemin, I., Beljonne, D., Blase, X.: J. Phys. Chem. Lett. **7**, 2814 (2016)

8. Li, J., D'Avino, G., Duchemin, I., Beljonne, D., Blase, X.: Phys. Rev. B **97**, 035108 (2018)
9. Yamada, K., Yanagisawa, S., Koganezawa, T., Mase, K., Sato, N., Yoshida, H.: Phys. Rev. B **97**, 245206 (2018)
10. Chen, Y., Tamblyn, I., Quek, S.Y.: J. Phys. Chem. C **121**, 13125 (2017)
11. Tamblyn, I., Darancet, P., Quek, S.Y., Bonev, S.A., Neaton, J.B.: Phys. Rev. B, **84**, 201402 (2011)
12. Quek, S.Y., Venkataraman, L., Choi, H.J., Louie, S.G., Hybertsen, M.S., Neaton, J.B.: Nano Lett. **7**, 3477 (2007)
13. Egger, D.A., Liu, Z.-F., Neaton, J.B., Kronik, L.: Nano Lett. **15**, 2448 (2015)
14. Aulbur, W.G., Jönsson, L., Wilkins, J.W.: In: Ehrenreich H., Spaepen F. (eds.) Quasiparticle Calculations in Solids. Solid State Physics, vol. 54, p. 1. Academic, Cambridge (2000)
15. Rieger, M.M., Steinbeck, L., White, I.D., Rojas, H.N., Godby, R.W.: Comput. Phys. Commun. **117**, 211 (1999)
16. Steinbeck, L., Rubio, A., Reining, L., Torrent, M., White, I.D., Godby, R.W.: Comput. Phys. Commun. **125**, 105 (2000)
17. Freysoldt, C., Eggert, P., Rinke, P., Schindlmayr, A., Godby, R.W., Scheffler, M.: Comput. Phys. Commun. **176**, 1 (2007)
18. Yanagisawa, S., Morikawa, Y., Schindlmayr, A.: Jpn. J. Appl. Phys. **53**, 05FY02 (2014)
19. Yanagisawa, S., Yamauchi, K., Inaoka, T., Oguchi, T., Hamada, I.: Phys. Rev. B **90**, 245141 (2014)
20. Yanagisawa, S., Hatada, S., Morikawa, Y.: J. Chin. Chem. Soc. **63**, 513 (2016)
21. Yanagisawa, S., Hamada, I.: J. Appl. Phys. **121**, 045501 (2017)
22. Rangel, T., Berland, K., Sharifzadeh, S., Brown-Altvater, F., Lee, K., Hyldgaard, P., Kronik, L., Neaton, J.B.: Phys. Rev. B **93**, 115206 (2016)

Printed in the United States
by Baker & Taylor Publisher Services